Water
Resources
and
Environmental
Sustainability

Water Resources and Environmental Sustainability

Gangadhar Banerjee
Sarda Ganguly

London • New Delhi

MV Learning
A Viva Books imprint

3, Henrietta Street
London WC2E 8LU
UK

4737/23, Ansari Road,
Daryaganj, New Delhi 110 002
India

ISBN: 978-93-87692-85-5

Printed and bound in India.

Dedicated to Revered Guru Maharaj Swami Bhuteshananda

This book is dedicated to revered Guru Maharaj Swami Bhuteshananda, the 12th President of Ramakrishna Math and Ramakrishna Mission, Belur Math, Howrah, West Bengal, India. His blessings were instrumental in completing the long journey. Let us pray to Guru Maharaj that His choicest blessings shower to all of us so that we may proceed towards the right direction.

Dedicated to My Parents and Grandparents

This book is also dedicated to my loving and affectionate parents and my daughter's grandparents Dr Indranarayan Banerjee and Chhayarani Banerjee, who left the material world for attaining eternal peace. Their absence created vacuum in our lives. They were instrumental in encouraging us to cultivate our inner instinct and devote energy, time and intelligence for doing substantial work especially in research for the new generation. Their blessings inspired us to complete the 24th publication in the series.

Contents

Preface

Water Resources and Environmental Sustainability is a major concern among the planners, policy-makers, environmentalists, scientists, economists, sociologists, financial institutions, non-governmental organizations, people and other stakeholders who are the key functionaries in the development process of the economy.

Water, a key driver of economic and social development, is only one of a number of vital natural resources. It is imperative that water issues are not considered in isolation. The magnitude of variability, timing and duration of high and low supply of water are not predictable. This equates to unreliability of the resource, which poses great challenges to water managers in particular and to societies as a whole.

Most developed countries have, in large measure, artificially overcome natural variability by supply-side infrastructure to assure reliable supply and reduce risks, albeit at high cost and often with negative impacts on the environment and on human health and livelihoods.

Many less developed countries, and some developed countries, are now finding that supply-side solutions alone are not adequate to address the ever increasing demands for water from demographic, economic and climatic pressures; waste-water treatment, water recycling and demand management measures are being introduced to counter the challenges of inadequate supply. In addition to problems of water quantity, there are also problems of water quality. Pollution of water sources is posing major problems for water users as well as for maintaining natural ecosystems. Water for social and economic development is clearly linked to the Integrated Water Resource Management (IWRM) which focusses on the three E's namely: equity, economy and environment.

There is a huge literature on water and environmental sustainability. The present treaty addresses the issues involved in the event of water scarcity and its impact on human life, climate change, irrigation, power generation, and ecosystem and so on in other words "Environment/Ecology".

The book is divided into **four sections** incorporating a total of seventeen chapters.

Section one deals with "Water Resources Management Policy". In this section, there are **five chapters.** Chapter one discusses water in the 21st century, followed by massive water crisis with hope for water restoration. Chapter two discusses generic and regional specific issues of water crisis and its peripherals in Chapter three. Chapters four and five discuss the failure in water resources planning and water resources management policy: revisited respectively.

Section two refers to "Integrated Water Resource Management" which includes **four chapters**. Chapter six showcases the role of water in economic and social development, while Chapter seven analyses the objective assessment of freshwater resources. Various aspects of integrated water resource management and its implementation along with the water resources management and its efficiency plan have been attempted in Chapters eight and nine respectively.

Section three envisages "Water Vision: A Management Paradigm" which again incorporates **four chapters**. Chapter ten states that an improvement of 10 per cent increase in the present level of water use efficiency in irrigation systems can bring additional 1 m.ha area under irrigation from the existing irrigation capacities at a very moderate investment and the steps initiated in this regard. Further, detail analysis has been made on the water crisis agenda and the role of financial institutions in promoting of rural water resources in Chapters eleven and twelve respectively. Business and protection of natural world has also been documented in Chapter thirteen. It suggests that the push towards improving environmental sustainability create many new business opportunities. Industries in the environmental sector are now emerging and growing at a rapid rate. There are numerous market opportunities for entrepreneurs, new enterprises and existing businesses to capitalize on in relation to environmental sustainability.

Section four examines "Environmental Sustainability and Economic/Social Progress". To address this issue **four chapters** have been dovetailed. Chapter fourteen devotes the Indian government's policy towards environment, status of India's environment – air quality, water resources and quality, public health in both rural and urban India, protection of environment to union and state governments, land degradation and soil losses, solid wastes and hazardous chemicals, etc. Chapter fifteen thoroughly examines environmental stability. Attempt has been made to highlight social mobilization and the peoples' participation for sustainability along with the role of National Institution for Transforming India (NITI) Aayog in Chapters sixteen and seventeen respectively.

This book is the collection of various articles published by the authors from time to time in different journals. The articles so collected have been modified, edited, updated and included in the proposed book. Besides, various literatures on water, environment and ecology have been studied. Annual reports of different departments of Government of India, Reserve Bank of India, NABARD and RBI bulletins have been consulted. In addition, publications and occasional papers of the Reserve Bank of India, NABARD and other financial institutions have been considered for preparation of the book.

Dr Gangadhar Banerjee

Sarda Ganguly

Abbreviations

ADB	Asian Development Bank
AfDB	African Development Bank
AICs	Atal Incubation Centres
AIM	Atal Innovation Mission
AIQ	Average Incremental Costs
ATLs	Atal Tinkering Labs
AOFFP	Area Oriented Fuelwood and Fodder Plantation
BOD	Biochemical Oxygen Demand
BOO	Build-Own-Operate
BOL	Build-Own-Lease
BOLT	Build-Operate-Lease-Transfer
BOOT	Build-Own-Operate-Transfer
BPL	Below Poverty Line
BSI	Botanical Survey of India
CAAQMS	Continuous Ambient Air Quality Monitoring Stations
CAD	Command Area Development
CBD	Convention on Biological Diversity
CC	Climate Change
CETP	Common Effluent Treatment Plants
CPCB	Central Pollution Control Board
C&I	Criteria and Indicators
CSE	Centre for Science and Environment
CSD	Commission on Sustainable Development
CSS	Central Sector Schemes
CST	Chhatrapati Shivaji Terminus

DDMs	District Development Managers
DFID	Department for International Development
EAP	Externally Aided Project
EDI	Economic Development Institute
EIA	Environmental Impact Assessment
EMENA	Europe, the Middle East and North Africa
EU	European Union
FAO	Food and Agriculture Organization
FFPS	Fuelwood and Fodder Project Scheme
FRBM	Fiscal Responsibility and Budget Management
GAP	Ganga Action Plan
GCM	Global Climate Model
GDP	Gross Domestic Product
GHGs	Greenhouse Gases
GNP	Gross National Product
GOI	Government of India
GWP	Global Water Partnership
HSD	High Speed Diesel
IAEPS	Integrated Afforestation and Eco-Development Projects Scheme
ICFRE	Indian Council of Forest Research and Education
IDA	Island Development Authority
IEO	Independent Evaluation Office
IIT	Indian Institute of Technology
IMT	Irrigation Management Transfer
IRR	Internal Rate of Return
IWDP	Integrated Wasteland Development Projects
IWMI	International Water Management Institute
IWRM	Integrated Water Resources Management
JPoI	Johannesburg Plan of Implementation
LRMC	Long Run Marginal Cost
MDG	Millennium Development Goals
MIDS	Madras Institute of Development Studies

MoEF	Ministry of Environment & Forests
MPC	Maharashtra Pollution Control
MKVDC	Maharashtra Krishna Valley Development Corporation
NAEB	National Afforestation and Eco-Development Board
NDA	National Democratic Alliance
NDC	National Development Council
NDRC	National Development and Reform Commission
NEAP	National Environment Action Plan
NEERI	National Environmental Engineering Research Institute
NGO	Non-governmental Organization
NITI	National Institution for Transforming India
NLCP	National Lake Conservation Programme
NMNH	National Museum of Natural History
NRCP	National River Conservation Plan
NTFP	Non-Timber Forest Produce
NTFPS	Non-Timber Forest Produce Scheme
NWMP	National Water Management Project
OECD	Organisation for Economic Co-operation and Development
O&M	Operations and Maintenance
PEO	Programme Evaluation Organization
PIM	Participatory Irrigation Management
PPP	Public-Private Partnership
PRA	Participatory Rural Appraisal
PRSP	Poverty Reduction Strategy Paper
PRIs	Panchayat Raj Institutions
RIDF	Rural Infrastructural Development Fund
RRBs	Regional Rural Banks
SARFAESI	Securitization and Reconstruction of Financial Assets and Enforcement of Security Interest Act
SEBI	Securities and Exchange Board of India
SLNWDR	Sri Lanka National Water Development Report
SSNNL	Sardar Sarovar Narmada Nigam Ltd.

STPs	Sewage Treatment Plants
TF	Task Force
TVA	Tennessee Valley Project
UNEP-UCC	United Nations Environment Programme Collaborative Center (Copenhagen)
UN	United Nations
UN-DESA	United Nations Department for Economic and Social Affairs
UNCED	United Nations Conference on Environment and Development
UNDP	United Nations Development Programme
UNESCO	United Nations Educational Scientific and Cultural Programme
UNEP	United Nations Environment Programme
UNFCCC	United Nations Framework Convention on Climate Change
UNIDO	United Nations Industrial Development Organization
USAID	United States Agency for International Development
WAYU	Wind Augmentation and Air Purifying Unit
WHO	World Health Organization
WSSD	World Summit on Sustainable Development
WUA	Water Users' Association
WWAP	World Water Assessment Programme
WWDR	World Water Development Report
WRCP	Water Resource Consolidation Project
WRM	Water Resource Management
ZSI	Zoological Survey of India

List of Tables

List of Annexures

Section 1

Water Resources Management Policy

CHAPTER 1

Water in 21st Century: Emerging Global Issues

1.0 Introduction

Earth is the blue planet with water being one of the most plentiful natural substances in its environment. There is more than 1.4 billion cubic kilometres (km³) of water, enough to give every man, woman and child more than 230 million cubic metres (m³) each if we were to divide it evenly. However, more than 98 per cent of the world's water is salt water and we depend for our basic vital needs on freshwater. Most freshwater is locked in the polar ice caps. Less than one per cent of the earth's freshwater is accessible in lakes, rivers and groundwater aquifers. That vital one per cent of available freshwater is constantly in motion, either flowing in rivers, evaporating and moving around the globe as water vapour, falling from the sky as rain or snow, or filtering slowly through the earth to emerge somewhere else. It is a renewable resource on which we all completely depend. It is the genesis and continuing source of all life on earth.

1.2 The Accessible Water

The most accessible water is that which flows in river channels or is stored in freshwater lakes and reservoirs. The major portion of the water diverted for human needs is taken from this renewable, readily accessible part of the world's freshwater resources. Although the total volume of water conveyed annually by the world's rivers is about 43,000 km³, most of this occurs as floods the low river flows (base flows) make up only about 19,000 km³. Of this, about 12,500 km³ can be assessed and present levels of withdrawal account for about 4,000 km³. Withdrawals are anticipated to reach, 5,500 km³ per year by 2025.

1.3 Demand for Freshwater

The demand for freshwater increased by six-fold between 1900 and 1995, twice the rate of population of growth. The 1997 United Nations (UN) Comprehensive Assessment of Freshwater Resources of the World concluded that one-third of the world's population today already live in countries experiencing medium to high water stress. High water stress and unsustainable rates of withdrawal are already being experienced in Central and South Asia, where annual water withdrawals compared with available water resources are 50 per cent or more. The northern People's Republic of China (PRC) and Mongolia have medium stress conditions with 25 per cent water use. Although water stress computes at less than 10 per cent in Southeast Asia (including southern China) and the Pacific and is, therefore, considered to be low, this measure is highly distorted by seasonally high river flows. In the dry season, water scarcity will affect food security throughout Asia and the Pacific.

1.4 Expansion of the Global Population

The global population will expand to almost 8 billion by 2025. By then, more than 80 per cent of the world's population will be living in developing countries. The World Meteorological Organization estimates, assuming the renewable water resources will remain unchanged, that the number of countries facing water stress will increase from 29 to 34 in 2025. How these countries manage their water resources, and whether they can produce sufficient food for their growing populations while catering to their water needs and preserving natural environments, have important implications and imperatives. The poverty reduction strategy adopted in 1999 enjoined Asian Development Bank (ADB), at the policy level, to continue to support Governments in developing, in a participatory manner, master plans for effective management of critical natural resources, including water.

1.5 Competition of Water Use

Competition for water is increasing among different water uses, including water for ecological needs. In many developing countries, irregular and inequitably distributed supplies of piped water have a detrimental effect on the social and economic well-being of most of their citizens. Ironically, consumers in almost all countries are charged less for their water than it costs to provide. Hence, utilities are reluctant to connect new customers, because water prices are too low to allow them to recoup their investment. For the poor, access to even a rudimentary level of municipal water supply is frequently denied, and they may be constrained to use untreated water from highly unreliable sources. Water-borne diseases are causing immense suffering and loss of productivity, with the poor suffering disproportionately. Large cities in Asia are not equipped to offer their burgeoning population the water supply and sanitation services they require.

1.6 Withdrawals of Global Freshwater

Nearly 70 per cent of global freshwater withdrawals are directed towards agriculture, mainly for irrigation. By some estimates (UN 1997) annual irrigation water use will have to increase about 30 per cent above present use for annual crop production to double and meet global food requirements by 2025. Although irrigation will remain the dominant water use in developing countries, an increase of 30 per cent in irrigation withdrawals may not be possible if other essential human needs are to be met. Making irrigation more efficient will be necessary and unavoidable. The industry sector, which accounts for about 22 per cent of current freshwater withdrawals globally is likely to require an increasing share in all regions of the world in both absolute and relative terms. In developing countries, where 56 per cent of population will be living in urban areas by 2025, the share of water going towards domestic uses will also need to grow substantially.

1.7 Water Pollution and the Environment

Emerging Asia, published by Asian Development Bank (ADB) in 1997, identified water pollution as the most serious environmental problem facing the region. Water pollution exacerbates the problem of water scarcity at local and regional levels by reducing the amount of water available for productive purposes. Water pollution comes from many sources, including untreated sewage, chemicals discharges, and spillage of toxic materials, harmful products leached from land disposal sites, agricultural chemicals, salt from irrigation schemes, and atmospheric pollutants dissolved in rainwater. The direct disposal of domestic and industrial wastewater into watercourses is the major source of pollutants in developing countries.

In Asia and the Pacific, pollution is one of the most serious problems, affecting both surface water and groundwater bodies and leading to a tenacious persistence of such water-borne diseases as cholera, typhoid and hepatitis. Estimates of the increase in water pollution loads in high growth areas of Asia over the next decades are as high as 16 times for suspended solids, 17 times for total dissolved solids and 18 times for biological pollution loading. The impact of this can be seen from the following comparison: the combined volume of water used and water needed to dilute and flush pollutants is almost equal to the volume of accessible freshwater in the world's river systems. The development of freshwater resources for human uses has compromised natural ecosystems that depend on these resources for their continued integrity.

1.8 Water Stress in River Basin – Degree of Annual Water Use

Water stress for a river basin is defined in degree of annual water use (that is water withdrawn from a surface or groundwater source for human purposes) as a percentage of the total water resources available in that basin. Water stress for a country is the summation of water stress for all its river basins. Water stress begins when withdrawals of freshwater rise above 10 per cent of renewable resources. Medium to high stress translates as water use that exceeds 20 per cent of available water supply. Countries experience high water stress when the ratio of water use to supply exceeds 40 per cent. At such levels, their patterns of use may not be sustainable, and water scarcity is likely to become the limiting factor to economic growth.

1.9 Unrestricted Development of Surface and Ground Waters

Freshwater ecosystems, comprising lakes, rivers and wetlands, have already lost a greater proportion of species and habitant than land or ocean ecosystems. Unrestricted development of surface water and groundwater has altered the hydrologic cycle and threatens the natural functions of deltas and wetlands. Wetlands have been converted to croplands, and rivers that channelled water to estuaries and deltas have dried up.

The Arab Sea basin illustrates vividly the extent to which human intervention has affected the natural functioning of aquatic systems. Excessive diversion of water for irrigation has been reduced the flow of rivers entering the sea. Thus, the surface of the river has shrunk by 45 per cent and its volume by 70 per cent since 1970. A formerly flourishing fishing industry has collapsed, and major health problems are now associated with windblown toxic salts and contaminated residues. Diminished productive potential, loss of vegetation, increased health risks and irreversible desecration of aquatic biota are the sad legacy.

1.10 Doubled Edged Sword: Floods and Droughts

Floods and droughts have always been features of life on earth and have produced some of the worst natural disasters in recorded history. Due to inappropriate land use and land management practices, uncoordinated and rapid growth of urban areas, and loss of natural flood storage wetlands, floods are becoming more frequent.

According to the Office of the United Nations Disaster Relief Coordinator, flooding is the hazard that affects more people than any other. Associated damage to property is escalating. Concurrently, destruction of forest cover has altered the hydrologic cycle and reduced water retention in forest soils. Accompanying soil erosion has permanently stripped fertile topsoil from vast areas, leading to further degradation of river basins and threatening the basis for sustainable natural resource management. Global climate change will have unpredictable but potentially devastating consequences for the hydrologic cycle by changing the total amount of precipitation, its annual and seasonal distribution, the onset of snowmelt, the frequency and severity of floods and droughts, and the reliability of existing water supply reservoirs. According to the intergovernmental Panel on Climate Change, the frequency of droughts could rise by 50 per cent in certain parts of the world by 2050.

1.11 Geographical Variability in Water Resources

Asia has the lowest per capita availability of freshwater resources among the world's continents. The contrasts within the region are stark. Annual freshwater resources (in m^3 per capita) reach as high as 200,000 in Papua New Guinea and as low as 2,000 in parts of South Asia and the PRC, and are generally below 20,000 in Southeast Asia. The region's weather is largely governed by a monsoon climate, which created large seasonal variations in addition to spatial variation. The two most populous nations in the world, China and India, will have 1.5 billion and 1.4 billion people, respectively, by 2025, by which time the availability of freshwater will have dropped to 1,500 m^3 per capita in India and 1,800 m^3 in China.

1. 12 Groundwater Exploitation to Supplement Scarce Surface Water Resources

Many of developing countries depend heavily on groundwater exploitation to supplement scarce surface water resources. In Bangladesh, groundwater abstraction already represents 35 per cent of total annual water withdrawals; in India 32 per cent; in Pakistan, 30 per cent and in China, 11 per cent. Groundwater overuse and aquifer depletion are becoming serious problems in the intensively farmed areas of northern China, India and Pakistan. In heavily populated cities such as Bangkok, Jakarta and Manila, land is subsiding as groundwater is withdrawn to serve the needs of their growing urban populations and saltwater intrusion is rendering much of the groundwater unusable.

1.13 International Conflicts Over Waters

International conflicts over water are becoming more frequent as competition for available freshwater resources increases. There are 215 international rivers as well as about 300 groundwater basins and aquifers that are shared by several countries. Although many difficult issues remain to be resolved, the 1996 treaty signed by Bangladesh and India for managing flows in the Ganges-Brahmaputra system represents a major victory for rational approaches to shared water resources. Similarly, Kazakhstan, Kyrgyz Republic, Tajikistan, Turkmenistan and Uzbekistan in recognition of their common strategic, economic and environmental interests created in 1992 the interstate Coordinating Water Commission to facilitate water sharing and common solutions to related environmental issues. However, more than 70 water-related flash points have been identified, mainly in Africa, Middle East and Latin America.

1.14 Status of Eight Countries

Eight countries in Asia (Bangladesh, Cambodia, Kazakhstan, Pakistan, Tajikistan, Thailand, Uzbekistan and Vietnam) rely on international rivers to supply more than 30 per cent of their annual water resources. Four external (Bangladesh, Cambodia, Uzbekistan and Vietnam) rely on water from external sources for more than 65 per cent of their annual water resources. The reliability of water supplies in the face of such dependence is a key issue when seasonal variations, particularly droughts and EL Niño events, enter the equation. Unsustainable rates of groundwater extraction can only make matters worse. The impact of global climate change, which cannot be determined now, will be to increase the overall uncertainty within which water planners operate.

1.15 Heightened Awareness of Water Issues

Traditionally seen as limitless bounty, water has only recently been recognized as a scarce resource, and only since the 1950s have policy-makers begun to espouse the economic and environmental values of water. Since the 1970s, a series of international meetings addressed water issues, starting with the First UN Water Conference of Mar del Plata in March 1977. This was followed by others. Major international conferences have drawn attention to the serious condition of the globe's freshwater resources in the last decade.

A consensus is growing among scientists, water planners, governments, and civil society that new policies and approaches will have to be adopted within the next two decades to avoid calamity, and that supply, use and management of water resources will have to be integrated across sectors and between regions sharing the same source. The concept of fully Integrated Water Resource Management (IWRM) emerged from the Dublin and Rio conferences in 1992. The four guiding principles now referred to as the Dublin Principles are: (i) Freshwater is a finite and vulnerable resource, essential to sustain life, development and the environment; (ii) water development and management should be based on a participatory approach, involving users, planners and policy makers at all levels; (iii) women play a central part in providing, managing and safeguarding water; and (iv) water has an economic value in all its competing uses and should be recognized as an economic good.

1.16 Elements of Water Strategy

The Dublin Principles recognize that freshwater is an input to which every human has the right to claim an essential minimum amount, the amount necessary to sustain life and meet basic sanitation needs. For human survival, the absolute minimum daily water requirement is only about 5 litres per day, whereas the daily requirement for sanitation, bathing and cooking needs as well as for assuring survival, is about 50 litres per person (equivalent to about 20 m^3 per year). Despite concerted efforts made during the 1980s (the International Drinking Water and Sanitation Decade), even this minimal amount was not provided in 55 countries (representing close to one billion people) by 1990.

One in five people living today does not have access to safe drinking water, and half the world's population does not have adequate sanitation. This is most acute in Asia where most of the world's poor people live. Not surprisingly, water and sanitation related diseases are widespread and increasing. Almost 250 million cases are reported each year, with about 10 million deaths. A recent UN report notes that "at any given time, 50 per cent of the population in developing countries is suffering from water-related diseases caused either by infection, or indirectly by disease carrying organisms." The global imperative is to ensure that at least 95 per cent of human beings have safe water and sanitation by 2025 (World Water Council 1999).

1.17 Investments in Water Supply and Sanitation

Water supply and sanitation in investments often do not keep pace with population growth. An estimated 737 million people in rural areas and 93 million in urban areas still have no access to safe drinking water. Access to sanitation is denied to 1.74 billion in rural areas and 298 million in urban areas. This is a major human tragedy; provision of such services to all people should be one of the highest priorities of all governments.

At the 1992 Rio Earth Summit, the rights of all human beings to basic daily water requirements were expanded to include environmental water needs. This was reinforced in a statement issued by the UN in 1997: "it is essential for water planning to secure basic human and environmental needs for water and develop sustainable water strategies that address basic human needs, as well as preservation of ecosystems."

1.18 Investment in Human Capital

Not only are the poor more prone to the adverse impacts of unsafe drinking water and inadequate sanitation, but ADB's field surveys also consistently show that the poor spend disproportionately more of their income on potable water than more privileged sections of the community for whom piped water supplies are assured. For example, the poor in Manila pay as much as 10 per cent of their household income for a meagre quantity of poor quality water. While investments in human capital (education, health care, shelter and protection from the effects of natural disasters) are also required to break the cycle of poverty, the impact of poor quality drinking water and the lack of adequate sanitation are particularly strong and immediate. The policy imperative of this is quite clear.

1.19 Central Role of Women in Providing, Managing and Safeguarding Water

While the poor are disadvantaged in terms of access to the benefits of improved water supply and sanitation, poor women are in a particularly unhappy situation. The gender division of labour in many societies allocates to women the responsibility for collecting and storing water, caring for children, sick, cleaning and maintaining sanitation. The availability of a decent water supply and sanitation system goes a long way in improving the quality of life for poor women and their families. In many parts of the region, the arduous task of walking long distances over difficult terrain to fetch water falls to women, often with the help of their daughters. Women care for the sick, who are often children suffering diseases caused directly by contaminated water. Providing clean and dependable water close to the home can substantially reduce women's workloads, and free up time for women to engage in economic activities to improve household incomes. For girls, the time saved can be used to attend school. Hence, providing water supply and sanitation is pivotal improving both the social and economic status of women, while simultaneously addressing gender and poverty concerns. The central role that women play in providing, managing and safeguarding water is recognized in the third Dublin Principle.

1.20 Water for Food Production

A major problem to be resolved by 2025 is producing enough food for the anticipated population of 8 billion people. Economic development and changes in food preferences will exert strong demand for additional production and more varied food products. In 1998, the International Water Management Institute (IWMI) stated that in many parts of the world, water is becoming the single most important constraint to increased food production.

Table 1.1: Major International Conferences in the 1990s

Year	Name of the Conference
1990	Safe Water and Sanitation for the 1990s (United Nations Development Programme (UNDP), New Delhi): appealed for concerted action to ensure access for all to the basic human needs to safe drinking water and environmentally sound sanitation.
1991	A Strategy for Water Sector Capacity Building (UNDP, Delft): defined the basic elements of capacity building necessary to create and enabling environment in the water sector.
1992	International Conference on Water and Environment (UN, Dublin): set out the four principles of water resource management that came to be known as Dublin Principles.
1992	United Nations Conference on Environment and Development (UN, Rio de Janeiro): promoted integrated water resource management based on the perception of water as an integral part of the ecosystem, a natural resource, and a socio-economic good.
1997	First World Water Forum (Marrakech): recommended action to meet basic human needs for clean water and sanitation, establish effective mechanism for management of shared waters, preserve ecosystems, encourage efficient use of water, address gender equality issues in water use, and encourage partnerships between civil society and governments.
1998	Water and Sustainable Development (United Nations Education, Scientific and Cultural Organization (UNESCO) and the French Government: raised concerns about tendencies to focus on scarcity as the main water crisis while neglecting problems of poor water management and the proliferation of regional coordination issues.
1999	Fifth Joint International Conference on Hydrology (UNESCO) and the world Meteorological Organization): drew attention to the catastrophic consequences of water mismanagement on the poorer communities in developing countries.

1.21 Technical Solutions on Hand

Even when good technical solutions appear to be hand, they do not always produce the expected results, and the poorest, most vulnerable members of the community are among the worst affected. In Bangladesh, for example, the use of tubewells to raise shallow groundwater has been promoted by funding agencies to support intensive irrigation while also providing safe drinking water in rural areas. This gave a dramatic boost during the last three decades to agricultural production, bringing the prospect of food self-sufficiency within reach for the first time. However, the same water has recently been found to contain traces of naturally occurring arsenic. Arsenic build up in the body initially manifests itself through the appearance of skin diseases, and prolonged ingestion damages internal organs leading to cancer and death. About 20 million people are at risk. The Government, with the aid of non-governmental organizations (NGOs) and international agencies, has embarked on a nationwide programme to define site-specific countermeasures, but this may not be adequate to avoid suffering and loss of life for people who cannot afford alternative water supplies.

1.22 Strategic Plan in Integrated Water in Minor Irrigation (IWMI)

IWMI in its draft strategic plan for 2002-2005 (October 1999) noted that "the potential for expanding irrigated area is extremely limited." The UN has estimated the potential area for new irrigation as 45 million hectares (m.ha) worldwide, which could provide up to 21 per cent of the projected additional food needs, increases in yield and cropping intensity are expected to provide the rest. However, erosion, water longing and land degradation are reducing the area of irrigated land, and some of the most fertile and productive areas close to urban centres are being absorbed into urban sprawl. In Indonesia, about 200,000 hectares are lost each year to urban development in Java alone. About 20 per cent of the world's 250 m.ha of irrigated land are degraded to the point where crop yields are declining.

1.23 Findings on the Study of Rural Areas

A study of rural Asia in 1998 noted that, from 1966 to 1988, the real cost of new irrigation schemes increased by more than 150 per cent in South and Southeast Asia. Given the limited scope for expanding irrigation and the sharply increased cost of new irrigation schemes, the justification for investing in new irrigation grows steadily weaker. The future irrigation lies mainly in improving the efficiency of present irrigation schemes in terms of operational performance and water use, supported by the introduction of mechanisms to ensure financial sustainability. Expansion of irrigation, where possible, will need to be justified on criteria of cost effectiveness related to other uses. Water scarce regions need to plan for a future in which they may be able to achieve food self-sufficiency.

1.24 Techniques for Water Harvesting and Supplemental Irrigation

In a world where food security can no longer be assured by an ever-expanding irrigation sector, what possibilities exist for increasing food production? The future of agriculture will increasingly be linked to careful use of marginal areas. Techniques for water harvesting and supplemental irrigation have shown great promise for increasing crop yields, and many scientists believe that rain-fed areas offer the greatest potential for increasing grain production in the future. More research will be needed, and both assistance and encouragement should be provided to poor farmers to help overcome their reluctance to invest time and scarce resources into inherently risky, farming. Smallholder water management systems, where groups of farmer's finance relatively small water capture and distribution infrastructure, can eliminate much of the insecurity of rainfed agriculture without increasing stress on the available water resources. Crops with low water requirements should be selected, and technology employed to determine accurately the exact amounts of water needed at different stages of growth. Innovative techniques for precision irrigation will help to increase the productivity of water ensuring more crop per drop. ADB supports research in selecting appropriate crops for non-irrigated areas, for example, by providing regional technical assistance (approved in 1999) for collecting, conserving and using indigenous vegetables.

1.25 Water as a Finite and Economic Good

The limits of the world's freshwater resources have become all too apparent, even though in many of the world's regions; detailed data on the hydrologic cycle are not available. Inefficient use, often initiated and then reinforced by government subsidies, has become ingrained; and the attendant water rights, whether formal or informal, are jealously defended by the privileged users. Agriculture and manufacturing use the greatest share of the world's water. Irrigation is particularly voracious, accounting for up to 80 per cent of water demand in hot, dry regions.

The river basin constitutes the natural hydrologic unit within which users compete for the same resource; and water quality is modified in ways that affect its value to other users. Management of water resources must therefore be approached on a comprehensive basis within this hydrologic unit. Beyond the basic needs for human well-being and environmental renewal scarcity of water is largely an economic issue. This idea, that water has an economic value in all competing uses and should be recognized as an economic good, must underlie all efforts for rational water resource management.

Part of the value of water is reflected in costs of extraction and delivery to the users. As a minimum, users should pay these costs to ensure accountability and financial sustainability. In addition, the opportunity cost, representing the value of the resource to some other users, must be considered. And finally, there are costs related to the impact on the environment and the health effects of polluted water.

1.26 Trading Water as a Tradable Commodity

Trading water as a tradable commodity would help ensure greater efficiency and productivity in its use. However, important cultural concerns and complex issues exist regarding resource sustainability and natural habitat, which means that Government intervention, is needed in resource allocation and investment decisions. Governments should therefore establish the policy, legislative and regulatory frameworks for managing water supply and demand. Governments should also provide financing for large water projects – dams, large-scale irrigation, flood control – for which private financing may not be readily available. They should also intervene, directly or indirectly to ensure that water resources are used in the most beneficial way for the greater society.

Allocations frequently become locked, however, into what are clearly low return uses (e.g., irrigation), when new projects are required to meet priority high-return needs (e.g., cities and industries). As the readily accessible water resources become committed, the costs of new projects can rise rapidly, resulting in high economic costs relative to the alternative of reallocating existing supplies. Even if countries are willing to incur the subsidies inherent in such solutions for instance, to meet social, political or environmental objectives the full burden of these subsidies is seldom transparent, large costs may inadvertently be incurred because of inefficient resource allocations resulting from such decisions.

1.27 Development to Water Resource Management

The past century has seen enormous changes in the way society conducts the business of economic development, food production and trade. Concurrently, and especially in the latter part of the century there has been an explosion in the construction of large projects for water storage, flood control irrigation and hydropower. These were conceived and realized in an atmosphere of challenge: how to tame nature to serve the needs of humanity.

1.28 Irrigated Area and Withdrawals of Freshwater

The limits to the scale of the projects were set by the ingenuity of engineering solutions. The driving forces were population growth, food security and industrial development. According to the UN Food and Agriculture Organization (FAO), irrigated area grew from 50–250 m.ha in the last century, and withdrawals of freshwater increased from 500 to about 4,000 km^3 per year.

Most large-scale projects have been financed by governments, and governments have naturally assumed responsibility for their management. The absence of private investors reflects not only the scale of the investments required, but also the fact that for some of these projects there were political objectives: for example, to encourage development in remote areas or to distribute development funds among regions. In many cases, it was assumed that users would repay the investment costs through water and other charges.

1.29 Water Resource Development to Supply and Demand Management

This did not always happen. The repayment obligations have been eased and the cost of providing the services has frequently become institutionalized as a direct subsidy. Water planners and developers have always worked from projections based on population growth, industrial and agricultural production, and level of economic and social development to determine demand, and hence to formulate engineering solutions to provide the appropriate freshwater supply. However, because of natural resource constraints and the accumulating adverse environmental impact of past projects, changes are beginning to be made in the way planners approach the problems of water supply.

This is evident as a discernible shift from water resource development towards supply and demand management. The tightening fiscal environment, recent financial crisis, and reduction in the potential for developing additional surface water and groundwater supplies have added impetus to this shift in the last decades. In addition, people the world over now place a higher value on maintaining the ecological function of freshwater ecosystems. There is also growing public pressure for the costs and benefits of water development projects to be shared more equitably and prudently and for investments to be directed towards satisfying basic human needs rather than benefitting elite groups at a high cost to the community at large. There is a heightened awareness of the issues relating to large dams, which are also relevant to other large-scale engineering solutions.

1.30 Public Goods to Priced Commodity

Improving the efficiency of water use is indispensable. In the United States, contrary to all expectations, total water use has declined by 10 per cent since 1980, even with population growth and a continued increase in economic wealth. Industrial use has declined by 40 per cent from a peak in 1970, while industrial output and productivity have both increased. Similarly, in Japan, where industrial output has soared since the 1970s, total industrial water use has fallen by 25 per cent. These reductions have been achieved through technological improvements (using less water to produce the same goods) and a change in the composition of industries making up the sector. The potential to reduce industrial use through further innovation, improved technology and cost incentives by 20-30 per cent. Comparable saving is possible in developing countries.

Residential water use, although only a small part (about 10 per cent) of total water use, can be reduced without sacrificing living standards. Readily available means include improving the efficiency of household appliances, better pricing structures, use of recycled water for certain applications and especially reducing unaccounted loss for water due to leaks and non-metered connections in aging distribution networks. In many cities, such as Dhaka, Jakarta and Manila, non-revenue water exceeds 50 per cent of water use.

1.31 Future Water Use of Human Needs is Irrigation

The single largest variable in future water use of human needs is irrigation. According to the UN's Economic and Social Commission for Asia and the Pacific, irrigation in Asia and the Pacific accounts for 80 per cent of total withdrawals compared with 70 per cent globally. By far the largest share of investments in agriculture during the green revolution era went into irrigation channels. These were often – and remain – heavily subsidized. The adopted technology was generally at the lowest end of the scale: as much as 60 per cent of the water is lost through leakage and evaporation before it even reaches the crop and an additional 20 per cent may be lost on the field. There are few incentives for the service providers or the farmers to improve the efficiency of water delivery and use in such schemes where water is free or priced well below its cost.

1.32 Improvements in Surface Canal Systems Gains in Efficiency

Agriculture's contribution to national income is declining in all developing countries. The agriculture sector is, therefore, coming under increasing pressure to release water to meet other, more productive needs. For producing high-value crops in water scarce areas, new irrigation techniques have been shown to be highly efficient and cost effective. Even simple improvements in surface canal systems, which are used almost exclusively in developing countries, can lead to impressive gains in efficiency. Efforts to increase efficiency in water use could, however, have serious impact on poor farmers, who may not be able to finance technological improvements. Hence, special assistance may be required to help poor farmers move up the technological ladder. Potentially greater saving can be achieved in delivering water, and the incentives for such improvements should be structured in such a way that the major beneficiaries (those who will avail of the "saved" water) contribute their share of the costs.

1.33 Expansion of Groundwater Irrigation by Using Public Funding

Groundwater irrigation presents a special case of too much of a good thing. Rapid expansion of groundwater irrigation during the last two decades, initiated using public funding but now largely driven by private investment, provided remarkable increases in yield, productivity and area of irrigated crops in parts of rural Bangladesh, China, India, Indonesia and Pakistan. However, unregulated extraction over vast areas has caused extensive and rapid lowering of the water table and, in coastal areas, contributed to salt water intrusion. In other parts, overwatering (combined with inadequate drainage) is bringing the water table dangerously close to the ground surface, rendering the surface saline and unusable. The productivity of some areas is now so threatened that large investments will be needed to avert complete collapse of the resource base.

1.34 Future Water Resource Projects

New projects for dams, water storage, irrigation, drainage, flood protection and water supply will continue to be needed in many countries where the basic water requirements for people have not yet been met. Sustainability criteria will predominate in decision making and emphasis will be given to environmental and social values. Increasingly, these projects will be financed with private sector participation where possible, and a wider range of stakeholders will be invited to participate in the process. Before deciding to invest in new storage and conveyance infrastructure, water planners will consider using existing infrastructure to meet the demands. These could be met through reallocation of the available water among users, taking advantage of the greater efficiency offered by improved technology and the opportunities for recycling water. Major obstacles to the rational reallocation of water among users, however, are legal and regulatory constraints on water transfer. In many countries, the complex systems of water rights inhibit the free movement of water as an economic good. An additional constraint is the lack of detailed understanding of the actual amount of water needed for various processes.

1.35 The Potential in Greater Use of Flood Damage Insurance

Because of their scale and the need to safeguard national and regional concerns above local interest, flood control and flood protection projects represents a special case in which private investment is unlikely to displace Government funding. However, the desire to secure higher levels of flood protection must be balanced against the effectiveness of non-structural alternatives (such as planning and building controls, enhancing wetlands, providing means for evacuating persons and livestock, flood-proofing of essential infrastructure, and improving flood warning systems), which are less expensive and which do not disturb the river system and its aquatic ecology. The potential exists in most countries for greater use of flood damage insurance. Properly managed, this would avoid the cost escalation of disaster rehabilitations and flood protection. Modern approaches emphasize balanced structural and non-structural measures within an integrated and comprehensive plan for management of natural resources in the river basin.

1.36 Water Scenario Across Asia and the Pacific

In terms of human needs, water availability is highly variable across Asia and the Pacific. In Singapore, affordable high quality water is available to all, 24 hours a day. In rural Nepal, fetching water for basic needs occupies up to four hours a day. Most people in the region do not have access in their homes to a 24-hour supply and are forced to boil or filter the water they obtain to make it potable. In urban areas, unaccounted for water average 35 per cent of production. Leakage (especially from house connections) probably accounts for half of this. Illegal connections, in adequate metering and slack metre reading account for the rest. Where tariffs are too low, excessive water consumption (more than 150 litres per capita per day) is common. By contrast, power bills are normally about four times those for water. Low tariffs mean that utilities are always struggling with financial viability and cannot contribute to capital investments.

1.36.1 Existence of Sewerage

It exists for less than 5 per cent of our regional population, and only about 20 per cent have onsite septic tanks. Basic latrines are available for about 50 per cent, but as many as 25 per cent have no formal sanitation at all. In urban areas, building control is lax and industries are often allowed to discharge effluents without treatment. Competition for water has become intense and, because prior claim to large portion of the resource has often been established (particularly by irrigators), urban suppliers are obliged to tap sources remote from the users. An example is the $400 m Melamchi Water Supply project in Nepal, which will draw water from three river basins outside the Kathmandu Valley to serve urban areas within the valley.

1.36.2 Problems Faced

The three main problems facing the sector are financial sustainability, water resources availability and equitable access. Planning for the long term is now critical. Water rights for domestic and industrial water supplies should be secured for at least 50 years. Tariffs need to be set to reflect the financial cost (and preferably the economic costs) of water. For example, in the water-scare Maldives, consumers in the capital of Malé pay the equivalent of $5 per cubic metre for desalinated piped water. Distortions in tariffs, where one part of a community cross-subsidizes another, need to be smoothed out, and all schemes should make adequate supplies available in poor areas. The poor can, and are willing, to pay for water.

1.36.3 Efforts in Rural Areas

In rural areas, special efforts are needed to reduce the distance of water supplies wherever possible and to encourage conservation approaches, such as rainwater harvesting. Based on ADB's evaluation of many water supply and sanitation projects, it is essential to include complementary education in hygiene to derive the full health benefits of improvements in infrastructure. Privatization of urban water supplies has not so far achieved a remarkably high success rate. Independent regulatory bodies are needed to reduce political interference and ensure accountable management and efficient delivery of water.

1.36.4 Future Challenges

The challenges in future are to: (i) open up competition, (ii) allow domestic privatization, (iii) allow existing utilities to operate with transparent cost recovery policies and independent regulatory bodies, (iv) greatly increase tariffs to affordable limits, and (v) introduce performance benchmarking in all utilities. An urgent need exists to reduce non-revenue water.

1.36.5 Flood Protection

For flood protection and all future projects using and controlling freshwater resources, the operational guidelines and procedures need to be adjusted to account for greater variability in climate because of global warming. This could add considerably to their costs.

1.36.6 Irrigation and Drainage

Irrigation and drainage projects have accounted for about 10 per cent of the total lending of the international financing agencies, and such financing has been the most important factor behind the rapid expansion in irrigated agriculture since the 1960s. In many cases, the irrigation schemes have performed well below expectations. Yet, the same agencies have repeatedly supported programmes for rehabilitation and improvement of physical infrastructure, often ignoring the institutional and statutory obstacles to more responsive service provision and sustainable operation. In response to growing competition for the available resources, both financial and physical, many countries are now tackling the challenge of policy and institutional reforms to achieve integrated water resource management and sustainable operation and maintenance of irrigation schemes needs to be given to the users.

1.37 Current Global Commitment of Financial Resources

The current global commitment of financial resources for all water-related infrastructure is estimated to be $80 billion annually. For water supply and sanitation alone, an annual investment of about $70 billion would be required over the next 10 years. Although official development assistance will continue to provide an important part of the necessary resources, the private sector will be called on increasingly.

1.38 Lessons Learned

ADB has implemented 437 water-related projects, for which financing totalling $15.7 billion has been provided. Evaluation studies show 51 per cent of these projects were generally successful, but 11 per cent were unsuccessful. Success was evaluated by a variety of indexes including economic internal rate of return many of which are affected by external factors, such as changes in the economic environment. For instance, declining real rice prices over the long term have adversely affected the economic evaluation of many irrigation projects.

1.39 Manila Case Study

1.39.1 Water and Poverty

Winnie lives in Block A of the Kabusig, Floodway, Cainta, Metro Manila. She is 34 and earns $162 per month as a domestic helper. Her husband a messenger earns $138 per month. They support a family of seven, which includes Winnie's mother-in-law and four children. They rent a 20-square metre room and share a kitchen and toilet with another family. With monthly expenses of $125 for food, $50 for transport, and $38 for rent, there is little left to cover costs of power, water gas, medicines and schooling. Water costs Winnie $20 per month, or 7 per cent of their household income. She used to pay $12.50 per month for a metered piped supply from a deep tubewell operated by a private contractor. However, the supply was only from a deep tubewell operated by a private contractor. However, the supply was only for one hour twice a day. She paid another $7.50 per month for drinking water purchased by the confessional piped supply.

1.39.2 Trouble with Sources of Water

Recently, there was trouble with both sources of water at the same time. The deep tube-well closed due to pump problems. Diarrhea and typhoid broke out in the neighbourhood. One of Winnie's boys had to be hospitalized. Now Winnie has a connection to a more distant deep tubewell and is selling water (not fit for drinking) by the container to five of her neighbours who do not have water. At $1.25 for the container to first 10 cubic metre (m^3) and $0.40 per m^3 thereafter, she fears she may not have collected enough to pay the excess charges at the end of the month as well as the installments on the $50 connection fee. The message brought home by Winnie's case study is that the poor can and do pay for water. Local governments should ensure that piped water supplies reach the poorest areas, and that the poor are assisted to make use of such supplies.

1.39.3 ADB's Analysis

An analysis of ADB's water operation shows positive trends for such concerns as the incorporation of social and environmental dimensions, increased water user responsibility and water use efficiency, cost recovery, institutional straightening, quality control, and monitoring arrangements. ADB's water projects, however, tended to be identified, processed, administered and evaluated within their sub-sector context, reflecting the fragmented approach to planning and implementing water projects in most developing countries. For example, legal aspects of water allocation have been addressed in less than one quarter of approved projects, and only one-third of the projects included water conservation measures. This tends to confirm that ADB's water loans have, in the past, focussed largely on improving water services (supply side solutions) in a subsector context, and that relatively few have addressed water resource issues, including water scarcity and efficient allocation of water between different uses.

The striking lesson from water-related projects is that, as competition for water increases, a more comprehensive and integrated approach to water operations is needed to encompass goals of social welfare, environmental integrity and economic productivity.

1.40 The Large Dam Debate

Reservoirs created by dams are essential for supplying water for human needs. They conserve water that would otherwise flow out of the river basin, and thereby enable release of water when river flows are insufficient. They enable the development of towns, industries and irrigation with all their economic and social benefits. They also provide hydroelectric power, flood mitigation and recreational facilities. Water can be released to maintain environmental flows, dilute pollutants and flush sediments out of the lower river reaches, thereby promoting healthier conditions and improving navigation.

1.41 Environmental and Social Impacts

Until the early 1980s, systematic evaluation of their environmental and social impacts was not mandatory. Such impacts are frequently serious but difficult to predict and quantify. The displacement of people to make way for the construction of dams and their reservoirs can cause great suffering and social dislocation. The negative ecological impacts can extend upstream into the reservoir and downstream to the sea. There is now growing opposition in most countries to new large dams, and several projects have recently been cancelled due to public opposition. There is also a stricter regulatory framework for such projects. International non-government organizations have played a role in fostering independent scrutiny of large dam projects, and they have emboldened the affected communities to seek a greater role in decisions that impact directly on their lives. The good dam sites (and many not so good sites) have already been used, and strict environmental and social conditions are now imposed.

1.42 Establishment of Independent World Commission

In response to growing concern, the World Conservation Union and the World Bank established an independent World Commission on Dams in 1997 to review their development effectiveness and develop standards, criteria and guidelines to guide decision makers in planning, implementing and decommissioning dams. Key issues in three areas social, environmental and economic engineering were examined to work towards a new consensus on the role of large dams in sustainable development. In support of this initiative, ADB is undertaking regional recommendations on best practices for evaluating designing, constructing, operating, monitoring and decommissioning dam projects in Asia.

1.43 Positive Outcome

A positive outcome of the growing opposition to large dams is the impetus this has given to finding new ways of solving problems of water scarcity. As a first step, planners now look for ways of improving the efficiency of existing physical infrastructure and distribution systems, introducing more efficient industrial processes, reallocating available water among competing users, and finding innovative ways of recycling water.

1.44 Fundamental Actions to Achieve this are:

- Stakeholder participation in all stages of the project cycle.
- Attention to the complementary roles of the public and private sectors, recognition of the special contribution of women, and incorporation of economic instruments to improve allocation efficiency.
- Integration of pro-poor strategies into project formulation to ensure that services are extended to poor areas and those rights of access are assured for the poor and other disadvantaged groups.

- Strengthening of regulatory and control function to maximize opportunities for private sector participation in service delivery.

- Environmental protection and enhancement is an integral part of every new project, with each project being evaluated in the whole river basin context.

- Acquiescence of directly affected communities prior to committing investment funds.

A new generation of water projects with an integrated approach to supply and demand management has emerged. These incorporate fully the principles of integrated water resource management and build on country-specific analyses of water resource needs, constraints, and potential. The first such analysis was made with ADB assistance in Sri Lanka in 1993 and led to the formulation of a national water sector profile and reform action plan. This was followed by ADB-supported institutional strengthening and policy reforms, which will pave the way for new investments in water resource development.

Other examples include Lao People's Democratic Republic, Pakistan and Vietnam where ADB is now supporting policy reforms and capacity building for integrated water resource management. An ADB financed assessment of the water sector in China (concluded in 1999) helped formulate strategic initiatives and an action plan that reflect a shift from a sectoral focus towards a more integrated and comprehensive approach. Projects now being prepared in China are tackling traditional water resource problems in conjunction with biodiversity conservation and legislative changes for improved natural resource management.

1.45 Water Logging and Salinity in Pakistan

1.45.1 Pakistan's Indus Basin Irrigation System

It is the largest contiguous irrigation system in the world with 3 major dams, 19 barrages and 43 interlinked canal systems. This vast system is served by an equally large drainage network commanding about 6 m.ha. More than 400,000 tubewells (mostly privately owned) provide groundwater to supplement the surface canal supplies.

Because of inequitable water application and inadequate drainage, 38 per cent of the irrigated area is now waterlogged. In addition, irrigation adds more than 1.2 tons of salt per year to each ha. The salt is carried into the root zone where it reduces yields. The high rates of evaporation characteristic of the region's semi-arid climate have rendered 14 per cent of the surface too saline for use.

1.45.2 Water Logging and Salinity Problems

Recognizing the seriousness of the water logging and salinity problems, the Government began an extensive and costly reclamation programme in 1959. Originally focussed on providing surface and subsurface drainage, the programme later included canal remodelling and selective lining to reduce aquifer recharge. The Government's 1993 Drainage Sector

Environmental Assessment recommended measures for overcoming the water logging and salinity problems:

- Precluding further developments that would mobilizes salt from deep groundwater aquifers.

- Restricting irrigation to areas where existing drainage is adequate.

- Ceasing public investment for drainage where improvements could reasonably be carried out by the private sector.

- Restricting subsurface drainage interventions to areas affected by saline groundwater, and only where there is an environmentally acceptable means of eliminating drainage effluent.

- Concentrating surface drainage interventions to areas at risk of storm water damage.

- Giving special attention to beneficiary participation in both structural and non-structural interventions.

1.45.3 Structural Intervention

It is recognized that structural interventions to control water logging and salinity need to be complemented by agricultural strategies that promote the efficient use of water, beginning at the national level and continuing to the field level. Accordingly, in 1995 the government adopted a long-term strategy for institutional reforms in the water sector.

The role of the Government was redefined with the objective of phasing out subsidies for operation and maintenance of irrigation schemes within 10 years, and decentralizing management of irrigation and drainage services. Separate organizations will be established for operation and maintenance of each main canal system, and secondary irrigation and drainage systems will be transferred to farmer organizations. Implementation started in 1996 and is being supported by major international financiers, including ADB, World Bank and Japan Bank for international Cooperation. Despite such environmental problems, irrigated agriculture continues to support the country's economic development and provides livelihood for millions of families.

1.45.4 Evolving Water Policy

To avail of ADB assistance, governments will need to adopt national water policies, laws, institutional reform, sector coordination mechanisms and a national water action agenda.

1.46 Regional Cooperation in the Water Sector

1.46.1 International Cooperation

International cooperation need not be complex and controversial when it comes to exchanging information and experience in water sector policies and reforms. While circumstances are different in each country, there are enough common issues in the water sector that make such an exchange useful and cost effective. Following its regional water policy consultation in 1996, ADB has promoted subregional water resource cooperation in Southeast Asia and South Asia in collaboration with the Global Water Partnership. These resulted in subregional water partnerships being established. ADB's regional water policy consultations in Southeast Asia concluded that:

1. Water has become the critical natural resource in most countries of Asia and the Pacific.
2. National action programmes are needed to manage water resources and improve water services that will sustain human and economic development in each developing member country in the coming decades.
3. Governments should provide leadership, commitment, and a focus on principles to direct an effective water sector reform process in each country.
4. National water apex bodies should be formed to oversee sector reforms.
5. A range of modalities for river basin organizations exists, and such river basin organizations need to respond to demand and suit local conditions.
6. Water conservation requires supply and demand management, pricing, charging, public awareness and ecosystem maintenance.
7. ADB should target the water sector in its operations with a long-term perspective and through effective partnerships to catalyze investments in integrated water sector programme in the region.

In South Asia, regional consultations resolved the followings:

• Sustainability of water resources, institutions and financing is critical to poverty reduction.

• National water policies need to adopt cross-sectoral approaches and be practical and implementable.

• Water institutions need to be reformed to deal with cross-sectoral dimensions through approaches that involve stakeholders at all levels.

• Participatory planning and management need to focus on people's needs, equity, gender and accountability.

1.46.2 Concentrated Efforts

Financial incentives and regulation, together with concerted efforts to protect water quality, aquatic ecosystems and watersheds, was reinforced to improve the efficiency and sustainability of resource use. Stakeholder recognition and participation will be promoted, and the needs of women and vulnerable groups will be adequately considered in water projects. New partnerships between public, private, community and NGO stakeholders will be developed to ensure effective policy reform and environmentally sustainable, socially acceptable projects.

1.46.3 Financial and Policy Support

Implementing such reforms will require sustained financial and policy support, for which ADB has a comparative advantage because of its long experience of working with water agencies in the region. In addition, its cofinancing modalities and experience in catalyzing private investments provide a window for increasing support from other funding agencies.

1.46.4 Changes in Climate

Possible changes in climate are of particular concern in Asia and the Pacific where such phenomena as monsoons and tropical cyclones play such a large role. Regional studies are supported on the possible impacts of climate change and is assisting its developing member countries to develop national response strategies to help them cope with the greater climatic uncertainty. Comprehensive coastal zone management plans have been prepared for countries vulnerable to sea level changes, and national strategies of managing water resources under conditions of heightened uncertainty will form part of the policy agenda.

Making better use of Asia's shared rivers is an unfinished agenda with potentially large benefits to millions of poor people in the region. However, formulating agreements between sub-regions to enable equitable sharing of resources and better control of transboundary pollution has proven to be highly controversial and, in some cases, strongly divisive.

1.47 Conclusion

The hope for the future lies in doing for water productivity what the green revolution did for crop productivity. This "blue revolution" as it has been termed by various scientists and water planners, would dramatically improve the efficiency of freshwater use, particularly in agriculture. The revolution will begin with greater public awareness of the potential dangers of a business as usual approach; help create policies, strategies and incentives needed to establish integrated water resource management on a global basis; and culminate in the allocation of resources to effect the social, institutional and technological changes necessary for efficient water allocation and use.

The magnitude of the challenges to be faced in water resource management during the coming decades is enormous if the worst-case scenarios are to be avoided. Despite the difficult choices that must be made worldwide to ensure sustainable water use and management, there is some cause for optimism. Commitment to stricter environmental controls and their enforcement does help to maintain healthier ecological conditions and can restore the severely degraded environment. The efficiency of water use can be improved without sacrificing quality of life, and such improvements can alleviate, if not completely avoid, looming water crises. Adopting socially inclusive policies to spread the benefits of water resource development to the poor and other traditionally disadvantaged members of the community benefits society will improve the living conditions, health, social stability and opportunities for productive employment.

Note

1. A daily water supply of 300 litres per person is considered an appropriate design standard for modern urban water supply schemes.
2. A large part of these losses return to rivers through the drainage network and as groundwater seepage and is therefore not lost from the river system although quantity flow is often poor and limits this usefulness.

References

Shiklomanov. I. A. 1997, Assessment of Water Resources and Water Availability in the World. Report prepared for the Comprehensive Assessment of Freshwater Resources of the World. United Nations: St. Petersburg.

Gleick, Peter H. 1998, *The World's Water, 1998-1999: The Biennial on Freshwater Resources,* Washington, DC: Island Press.

United Nations Industrial Development Organization 1996, Global Assessment of the Use of Freshwater Resources for Industrial and Commercial Purposes. Industry, Sustainable Development and Water Programme Formulation, Technical Report, United Nations, New York.

CHAPTER 2

Water Crisis Is Massive but there Is Hope

2.0 Introduction

More than one in every six people in the world is water stressed, meaning that they do not have access to potable water. Those that are water stressed make up 1.1 billion people in the world and are living in developing countries. According to the Falkenmark Water Stress Indicator, a country or region is said to experience "water stress" when annual water supplies drop below 1,700 cubic metres per person per year. At levels between 1,700 and 1,000 cubic metres per person per year, periodic or limited water shortages can be expected. When a country is below 1,000 cubic metres per person per year, the country then faces water scarcity. In 2006, about 700 million people in 43 countries were living below the 1,700 cubic metres per person threshold. Water stress is ever intensifying in regions such as China, India, and Sub-Saharan Africa, which contains the largest number of water stressed countries of any region with almost one-fourth of the population living in a water stressed country. The world's most water stressed region is the Middle East with averages of 1,200 cubic metres of water per person. In China, more than 538 million people are living in a water-stressed region. Much of the water stressed population currently live in river basins where the usage of water resources greatly exceeds the renewal of the water source.

2.1 Water Is Life

Wherever they are, people need water to survive. Not only is the human body 60 per cent water, the resource is also essential for producing food, clothing, and computers, moving our waste stream, and keeping us and the environment healthy. Unfortunately, humans have proved to be inefficient water users. (The average hamburger takes 2,400 litres, or 630 gallons, of water to produce, and many water-intensive crops, such as cotton, are grown in arid regions.)

According to the United Nations, water use has grown at more than twice the rate of population increase in the last century. By 2025, an estimated 1.8 billion people will live in areas plagued by water scarcity, with two-thirds of the world's population living in water-stressed regions as a result of use, growth, and climate change. The challenge we face now is how to effectively conserve, manage, and distribute the water we have. National Geographic's Freshwater Website encourages to explore the local stories and global trends defining the world's water crisis. Learn where freshwater resources exist; how they are used; and how climate, technology, policy, and people play a role in both creating obstacles and finding solutions. Peruse the site to learn how one can make a difference by reducing one's water footprint and getting involved with local and global water conservation and advocacy efforts.

2.2 The Global Water Crisis

The global water crisis claims that 3.4 million lives each year. It is 3.4 million people with names, families, hopes and dreams. When confronted with this reality we must respond. This is not someone else's crisis, it is all of ours.

The water crisis is massive but there is hope. In many places water crisis exists, it is not because of physical lack of water, but the people do not have access to water that directly benefit them from water stress. The United Nations (UN) estimates that, of 1.4 billion cubic kilometres (1 quadrillion acre-feet) of water on Earth, just 200,000 cubic kilometres (162.1 billion acre-feet) represent fresh water available for human consumption.

2.3 Decline of Available Fresh Water – Changes in Climate

The popular opinion is that the cause of the total amount of available freshwater supply is decreasing because of climate change. Climate change has caused receding glaciers, reduced stream and river flow, and shrinking lakes and ponds. Many aquifers have been over-pumped and are not recharging quickly. Although the total fresh water supply is not used up, much has become polluted, salted, unsuitable or otherwise unavailable for drinking, industry and agriculture. To avoid a global water crisis, farmers will have to strive to increase productivity to meet growing demands for food, while industry and cities find ways to use water more efficiently.

2.4 Water Shortages Ties to Population Size More than Rainfall

A *New York Times* article, "Southeast Drought Study Ties Water Shortage to Population, Not Global Warming", summarizes the findings of Columbia University researcher on the subject of the droughts in the American Southeast between 2005 and 2007. The findings published in the *Journal of Climate* say that the water shortages resulted from population size more than rainfall. Census figures show that Georgia's population rose from 6.48 to 9.54 million between 1990 and 2007. After studying data from weather instruments, computer models and tree rings measurements which reflect rainfall, they found that the droughts were not unprecedented and result from normal climate patterns and random weather events. "Similar droughts unfolded over the last thousand years", the researchers wrote, "Regardless of climate change, they added, similar weather patterns can be expected regularly in the future, with similar results." As the temperature increases, rainfall in the Southeast will increase but because of evaporation the area may get even drier. The researchers concluded with a statement saying that any rainfall comes from complicated internal processes in the atmosphere and are very hard to predict because of the large amount of variables.

When there is not enough potable water for a given population, the threat of a *water crisis* is realized. The United Nations and other world organizations consider a variety of regions to have water crises a global concern. Other organizations, such as the Food and Agriculture Organization, argue that there are no water crises in such places, but steps must still be taken to avoid one.

2.5 Causes of Water Crisis

There are several principal manifestations of the water crisis. These are : (i) Inadequate access to safe drinking water for about 884 million people, (ii) Inadequate access to water for sanitation and waste disposal for 2.5 billion people (iii) Groundwater overdrafting (excessive use) leading to diminished agricultural yields, (iv) Overuse and pollution of water resources harming biodiversity and (v) Regional conflicts over scarce water resources sometimes resulting in warfare.

2.6 Unsafe Drinking Water, Inadequate Sanitation and Poor Hygiene

Waterborne diseases and the absence of sanitary domestic water are one of the leading causes of death worldwide. For children under age five, waterborne diseases are the leading cause of death. At any given time, half of the world's hospital beds are occupied by patients suffering from waterborne diseases. According to the World Bank, 88 per cent of all waterborne diseases are caused by unsafe drinking water, inadequate sanitation and poor hygiene.

Water is the underlying tenuous balance of safe water supply, but controllable factors such as the management and distribution of the water supply itself contribute to further scarcity.

A 2006 United Nations report focuses on issues of governance as the core of the water crisis, saying "There is enough water for everyone" and "Water insufficiency is often due to mismanagement, corruption, lack of appropriate institutions, bureaucratic inertia and a shortage of investment in both human capacity and physical infrastructure". Official data also shows a clear correlation between access to safe water and GDP per capita.

2.7 Lack of Property Rights, Government Regulations, Subsidies, Etc.

It has also been claimed, primarily by economists, that the water situation has occurred because of a lack of property rights, government regulations and subsidies in the water sector, causing prices to be too low and consumption too high.

Vegetation and wildlife are fundamentally dependent upon adequate freshwater resources. Marshes, bogs and riparian zones are more obviously dependent upon sustainable water supply, but forests and other upland ecosystems are equally at risk of significant productivity changes as water availability is diminished. In the case of wetlands, considerable area has been simply taken from wildlife use to feed and house the expanding human population. But other areas have suffered reduced productivity from gradual diminishing of freshwater inflow, as upstream sources are diverted for human use. In seven states of the U.S. over 80 per cent of all historic wetlands were filled by the 1980s, when Congress acted to create a "no net loss" of wetlands.

In Europe extensive loss of wetlands has also occurred with resulting loss of biodiversity. For example many bogs in Scotland have been developed or diminished through human population expansion. One example is the Portlethen Moss in Aberdeenshire.

2.8 Water Crisis by 2020

According to the California Department of Water Resources, if more supplies aren't found by 2020, the region will face a shortfall nearly as great as the amount consumed today. Los Angeles is a coastal desert able to support at most 1 million people on its own water; the Los Angeles basin now is the core of a megacity that spans 220 miles (350 km) from Santa Barbara to the Mexican border. The region's population is expected to reach 41 million by 2020, up from 28 million in 2009. The population of California continues to grow by more than 2 million a year and is expected to reach 75 million in 2030, up from 49 million in 2009. But water shortage is likely to surface well before then.

2.9 Water Crisis Leading Grain Deficit

Water deficits, which are already spurring heavy grain imports in numerous smaller countries, may soon do the same in larger countries, such as China and India. The water tables are falling in scores of countries (including Northern China, the US, and India) due to widespread over-pumping using powerful diesel and electric pumps. Other countries affected include Pakistan, Iran and Mexico. This will eventually lead to water scarcity and cutbacks in grain harvest. Even with the over-pumping of its aquifers, China is developing a grain deficit. When this happens, it will almost certainly drive grain prices upward. Most of the 3 billion people projected to be added worldwide by mid-century will be born in countries already experiencing water shortages. Unless population growth can be slowed quickly, it is feared that there may not be a practical non-violent or humane solution to the emerging world water shortage.

After China and India, there is a second tier of smaller countries with large water deficits – Algeria, Egypt, Iran, Mexico, and Pakistan. Four of these already import a large share of their grain. But with a population expanding by 4 million a year, it will also likely soon turn to the world market for grain.

2.10 Agricultural Crisis

Although food security has been significantly increased in the past thirty years, water withdrawals for irrigation represent 66 per cent of the total withdrawals and up to 90 per cent in arid regions, the other 34 per cent being used by domestic households (10 per cent), industry (20 per cent), or evaporated from reservoirs (4 per cent). (*Source:* Shiklomanov, 1999)

As the per capita use increases due to changes in lifestyle and as population increases as well, the proportion of water for human use is increasing. This, coupled with spatial and temporal variations in water availability, means that the water to produce food for human consumption, industrial processes and all the other uses is becoming scarce.

2.11 United Nations Climate Report

According to a UN climate report, the Himalayan glaciers that are the sources of Asia's biggest rivers – Ganges, Indus, Brahmaputra, Yangtze, Mekong, Salween and Yellow – could disappear by 2035 as temperatures rise. It was later revealed that the source used by the UN climate report actually stated 2350, not 2035. Approximately 2.4 billion people live in the drainage basin of the Himalayan rivers. India, China, Pakistan, Bangladesh, Nepal and Myanmar could experience floods followed by droughts in coming decades. In India alone, the Ganges provides water for drinking and farming for more than 500 million people. The west coast of North America, which gets much of its water from glaciers in mountain ranges such as the Rocky Mountains and Sierra Nevada, also would be affected.

2.12 Severe Ecological Damage for the whole Murray – Darling Basin

By far the largest part of Australia is desert or semi-arid lands commonly known as the outback. In June 2008 it became known that an expert panel had warned of long term, possibly irreversible, severe ecological damage for the whole Murray-Darling basin if it does not receive sufficient water by October. Water restrictions are currently in place in many regions and cities of Australia in response to chronic shortages resulting from drought. The Australian of the Year 2007, environmentalist Tim Flannery, predicted that unless it made drastic changes, Perth in Western Australia could become the world's first ghost metropolis, an abandoned city with no more water to sustain its population. However, Western Australia's dams reached 50 per cent capacity for the first time since 2000 as of September 2009. As a result, heavy rains have brought forth positive results for the region. Nonetheless, the following year, 2010, Perth suffered its second-driest winter on record and the water corporation tightened water restrictions for spring.

2.13 Physical and Economic Scarcity

Around one fifth of the world's population currently live in regions affected by Physical water scarcity, where there is inadequate water resources to meet a country's or regional demand, including the water needed to fulfil the demand of ecosystems to function effectively. Arid regions frequently suffer from physical water scarcity. It also occurs where water seems abundant but where resources are over-committed, such as when there is over development of hydraulic infrastructure for irrigation. Symptoms of physical water scarcity include environmental degradation and declining groundwater as well as other forms of exploitation or overuse.

Economic water scarcity is caused by a lack of investment in infrastructure or technology to draw water from rivers, aquifers or other water sources, or insufficient human capacity to satisfy the demand for water. One quarter of the world's population is affected by economic water scarcity. Symptoms of economic water scarcity include a lack of infrastructure, causing the people without reliable access to water to have to travel long distances to fetch water, that is often contaminated from rivers for domestic and agricultural uses. Large parts of Africa suffer from economic water scarcity; developing water infrastructure in those areas could therefore help to reduce poverty. Critical conditions often arise for economically poor and politically weak communities living in already dry environment.

2.14 Deforestation of Madagascar's Highland Plateau

On Madagascar's highland plateau, a massive transformation occurred that eliminated virtually all the heavily forested vegetation in the period 1970 to 2000. The slash and burn agriculture eliminated about 10 per cent of the total country's native biomass and converted it to a barren wasteland. These effects were from overpopulation and the necessity to feed poor indigenous peoples, but the adverse effects included widespread gully erosion that in turn produced heavily silted rivers that "run red" decades after the deforestation. This eliminated a large amount of usable fresh water and also destroyed much of the riverine ecosystems of several large west-flowing rivers. Several fish species have been driven to the edge of extinction and some, such as the disturbed Tokios coral reef formations in the Indian Ocean, are effectively lost. In October 2008, Peter Brabeck-Letmathe, the then chairman and former chief executive of Nestlé, warned that the production of biofuels will further deplete the world's water supply.

2.15 Inadequate Drinking Water – Crisis Impacts

There are many other countries of the world that are severely impacted with regard to human health and inadequate drinking water. The following is a partial list of some of the countries with significant populations (numerical population of affected population listed) whose only consumption is of contaminated water: Sudan 12.3 million, Venezuela 5.0 million, Ethiopia 2.7 million, Tunisia 2.1 million, Cuba 1.3 million.

Several world maps showing various aspects of the problem can be found in this graph article.

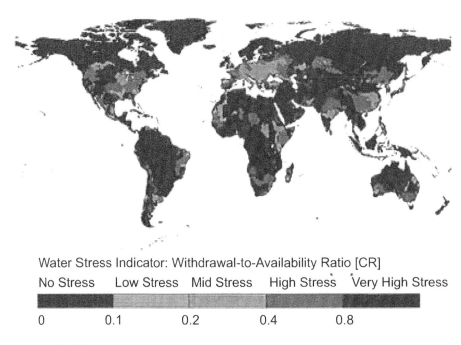

Water Stress Indicator: Withdrawal-to-Availability Ratio [CR]

No Stress   Low Stress   Mid Stress   High Stress   Very High Stress

0   0.1   0.2   0.4   0.8

Source: Water GAP 2.0

The United Nations Committee on Economic, Social and Cultural Rights established a foundation of five core attributes for water security. They declare that the human right to water entitles everyone to sufficient, safe, acceptable, physically accessible, and affordable water for personal and domestic use.

2.16 Conclusion

In order to address the effects of economic water scarcity, increasing access of safe, accessible drinking water was made an international development goal by the United Nations at the Millennium Summit in the year 2000. During this time, the Millennium Development Goals were drafted and eight goals were agreed upon by all 189 UN members. MDG 7 sets a target for reducing the proportion of the population without sustainable access to safe drinking water by half by the year 2015. This would mean that more than 600 million people would gain access to a safe source of drinking water.

According to *D+C (Development and Cooperation)* magazine, MDG 7 is still far from being reached as of 2012. Since national governments often cannot meet the infrastructure needs for growing populations on their own. According to the magazine, governments are receiving assistance from many NGOs that put forth their effort in assisting developing nations in the realm of water security.

CHAPTER 3

Water – Generic and Region Specific Issues

3.0 Back Drop

The World Bank has operations in 152 countries, aggregating them into four major regions. To formulate Bank policy with respect to water it is essential to distinguish between issues that are generic to the management of water as a resource, and others that are specific to individual regions. For example, the difficulty of making "correct" allocation decisions between different uses or users is generic to water around the globe, but the relative importance of irrigation varies by climatic zone and water availability. The generic issues are the ones for which the Bank needs to evolve institution-wide policy; the remaining issues can be left to the regional departments.

Fourteen generic issues are briefly described and discussed below; nine as broad sector-wide issues, and five as sub-sector issues.

3.1 Generic Issues of Concern for the World Bank

In the water planning community, there are several connections of 'water and... "Which are taken by many as paramount relationships that brook no questioning? As argued below, the Bank's Comprehensive Water Resources Management Policy Paper should take a close look at such sanctified linkages as water and health, water and food, water and equity, water scarcity, irrigation efficiency, and so forth, leaving no assumption unexamined.

3.2 Broad Sector-Wide Issues

3.2.1 International Conflicts over Water

Worldwide there are hundreds of rivers shared by different nations (54 just between India and Bangladesh). In many cases serious bilateral or multilateral conflicts over water have arisen. The current dispute between Turkey, Syria, and Iraq over the Euphrates is a good example of such situations. International law on transnational rivers is weak; it is essentially left up to the goodwill of the upstream riparian to settle problems amicably.

Several decades ago, the World Bank was a major player in one of the most successful conflict resolutions; the Indus Basin settlement between India and Pakistan.

The Bank should maintain its rule of not making loans for a project that is not acceptable to a country's neighbours, but it should also consider establishing a negotiating unit to help countries deal with these issues. In many cases the provision of a neutral corner and unbiased international expertise could make it politically easier for countries to explore their water conflicts with neighbours in a non-threatening situation, relatively free of inhibiting political limelight and implicit threats to national sovereignty. Although regrettably not fruitful to date, the Bank's work over many years on hydro-electricity conflicts between Nepal and India is an example of such an approach, and a credit to the Bank. Easing some of these disputes would open many new investment possibilities for constructive multi-purpose use of internationally shared river basins.

3.2.2 Linking the Water Sector to the National Economy

It is a paradox that although water resources have probably received more analytic attention than any other kind of public investment, there has been little attention paid to relating the water sector to inter-sectoral or macro allocation decisions. This lack of analysis should be of great concern in countries such as Brazil where 30 per cent of public investment is in the water sector. Other resource sectors such as energy have well developed methodologies to relate sectoral and macro plans. The development of reliable planning methodology to relate water sector plans to the overall macro development of a country is a generic problem that is of major significance to guiding investments and other Bank water activities. In a paper with Hurst (Hurst and Rogers, 1985) the author found in the literature only a few attempts to do this. Hurst and Rogers built an economy-wide model for Bangladesh which incorporated a detailed water sector and its macro-economic linkages. When the model was run, the optimal solution was radically different from the optimal solution predicted when the components were run separately. In the overall model, there was a strong bias against producing an export crop because:

The reason for this is that the growth of the non-agricultural sectors depends upon the import of raw materials and intermediate goods. If the food requirements of the rapidly growing population must be met with food-grain imports (for a given level of imports) less raw materials and intermediate materials can be imported. Consequently, non-agricultural sectors must grow at a slower rate, and to meet a reasonable level of national growth, the agricultural sector must grow rapidly.

While this model was never used in planning, it does indicate that quite different policy implications arise when the macro linkages are explicitly considered. Additional research in this area would help improve the quality of water investments in all sectors.

3.2.3 Water Sector Planning Methodology and Data Requirements

At the next level down, within the water sector there is a plethora of water planning tools. One of these has come to be called "multi-objective planning". Although conceptually satisfying, planning theoreticians have made the approach unnecessarily complicated with its own nomenclature and dedicated computer software. Multi-objective planning is based upon the concept of constrained optimization. One merely optimizes one objective, for example, national economic growth, while setting other objectives, such as environmental quality, as constraints upon the system. The Bank itself has already been a major contributor to the development of "multi-objective planning" through its modelling work in Mexico and Pakistan. The issues of equity, food self-sufficiency, and health discussed below are typical of the issues whose opportunity costs can be established by such methods. There is no doubt that this approach should be taken in any planning study. Whether specialized software or more general purpose models such as linear and non-linear programming or simulation models are used, the different versions of this tool yield essentially the same results, and like all analytic approaches should be employed with a great deal of caution.

3.2.4 Water Resources Influenced by Comprehensive Multi-Purpose River Basin Planning

The literature on water resources is also heavily influenced by the ideas of comprehensive multi-purpose river basin planning. This has historically been the approach taken in Western Europe and the United States. Although intellectually satisfying, it is not clear whether it is the best approach to practical problems. It is helpful for the physical aspects of rainfall and run-off but much less helpful from the point of view of political jurisdictions and economic markets. In the United States, it has since been abandoned by all major agencies involved in national water planning. Today agencies allow the specific problem to dictate the unit of analysis. With increasing concern for broader problems such as environmental quality, the idea of analyzing "problem-sheds" rather than river basins is gaining ground. A recent book by Major and Schwarz shows how river basins can be building blocks in larger problem-sheds. The Bank will have to examine closely the types of issues faced by its clients before recommending a planning approach.

3.2.5 Little Planning Applied Across Water Uses in the Economy

At the practical level, however, little planning of any kind seems to have been applied consistently across water uses in the economy. For example, in many places the uses of water are still examined separately on a project by project basis. Thus, in Algeria it is possible to find examples of carefully planned irrigation projects alongside carefully planned urban water supply projects, with no way of reconciling the conflicting demands that both have for the same water supply. In Brazil, there was surprise when the introduction of a tariff on wastewater discharges led to large decreases in revenue to the water supply utility, almost leading to its bankruptcy. In these cases, the unit of analysis was not large enough to capture the external effects of policies dealing with individual components.

3.2.6 Appropriate Models of the Water Sector

There is a need to ensure that appropriate models of the water sector be included in any project appraisal. The unit of analysis must be large enough to ensure that most of the relevant externalities are captured in the analysis. In irrigation projects, ground and surface water as well as drainage must be analyzed as one unit. Urban water supply cannot be divorced from wastewater disposal and its downstream impacts. Water diversions from rivers must consider alternative in-stream uses of the same water for navigation, fisheries, downstream users, and the maintenance of stream ecosystems. Well established methods using mathematical programming and simulation models should be required as a prerequisite for project appraisal in such cases.

3.2.7 Use of Quality Data in Policy Study

In any policy study the quality of the data used determines to a large extent what conclusions can be reliably reached. However, there is a mind-set, that always looks for more and more data and more and more complete coverage of a region or basin. Geographical information systems (GIS) are currently very popular. Some of the practical problems encountered in the use of such systems should be examined. For instance, many of the current systems have been devised with little or no idea about how the information is to be used. This results in over-collection of some types of data and under-collection of others. The experiences of GIS users to date have not been uniformly happy. Most systems concentrate solely upon physical and scientific data, and there is little or no coverage of important economic facts. Although many of the economic data do not need to be collected according to detailed geographic distributions, the economic realities may be as important as the physical data that do, and may be critical to understanding the meaning of the physical givens. More attention should also be paid to the statistical variability of the data. Currently available methods to establish statistical reliability should be insisted upon in all data analyses and be carried over into the analyses of investment decisions.

3.2.8 Planning Methods Elucidate the Uncertainty regarding the Decision

One issue that ties the methodology and the data together is the issue of "uncertainty". Not only is it important to develop and use reliable data but it is also imperative that the planning methods used elucidate the uncertainty regarding the planning decisions. All too often in the water resources area the bulk of the analysis is aimed at the consequences of hydrological uncertainty when, as demonstrated by James, Bower, and Matalas (1969), the largest part of the uncertainty in the implied decisions stems from the economic parameters which typically receive little stochastic evaluation.

Here serious effort should be made to show the extent of realistic expectations about the data that may reasonably be gathered, and the levels of reliability that can be achieved using these data.

3.2.9 The Political Economy of Pricing

Apart from certain aspects of religious and health significance, most people in everyday use do not regard water as an end. It is a commodity like "ships and shoes and sealing wax," consumed directly or used as an input to other processes. Empirical observation shows that in Boston as well as in Beijing, if the price of water is increased, consumers will use less of it. In other words, there is a "willingness-to-pay" for water that is not an abstract economic concept depending on elaborate theories of private property, but rather is a reliable behavioural trait of consumers of the products. The same behaviour can be expected in Kansas and Kathmandu – and is a good basis for assessing the economic demand for water. (Water has a downward sloping demand curve.) We can expect human actions to meet demand to be similar in widely different locations. The most accessible water for human use is the cheapest to develop, the next most accessible costs a little more, the next yet a little bit more, and so forth. Very soon one observes an upwardly sloping "marginal cost" curve that depends upon nature, not upon theory. Based upon these two observed phenomena of willingness to pay and increasing marginal cost curves, a good set of practical rules can be derived for helping to decide upon how much to invest in developing a particular water supply.

3.2.10 Misconception of Water as Public Goods

Although most people perceive water in this way at an individual level, the same individuals treat water differently when they gather collectively to make decisions about future uses of water. This contradiction may be due to the common social misperception of water as a "pure public good" which belongs to "everyone" at the same time, with a right of access for all. As a result, it is customary to treat water as a free good when it is in fact anything but free. People come to expect that it is their right to take (and to waste) as much water as they want (Baumol and Oates, 1979).

3.2.11 Water as Pure Public or Private Good

There is clearly a paradox here, since it is obvious that for many uses (for example, irrigation and municipal supply) water has all the properties of an exclusive economic good–just the opposite of a "public good". Adam Smith was one of the first to define a "pure public good": once the good is provided, it is not possible to exclude anyone who wants to take advantage of a public good from using it, and the consumption of the public good by any one consumer does not impede the consumption by any other potential consumer. At the other extreme, a "pure private good", such as food purchased from a market, can be and usually is the exclusive property of the owning individual; his or her consumption of the food absolutely prevents anyone else from consuming it. Water, like many other goods, falls somewhere between these two extremes. It has "pure public good" aspects when it is left in a scenic river, but even then too many people using the scenic river will destroy some of its value for other participants. It has "exclusive economic good" features when it is evaporated by farmers irrigating their crops for profit in a market setting.

It is the in-between cases which tend to predominate in the academic works on water and which confuse the issue. The question as to who has access to water becomes very important in determining how water is defined, and this definition in turn hinges upon the different doctrines governing water rights. To use water more efficiently it may be necessary to codify the water rights.

3.2.12 Market a System of Property Rights to Scarce Goods

Markets are based upon a system of property rights to scarce goods and the right to exclude other users of the resource. A private water market can only exist if property rights are secure and can be transferred. However, water law and water rights differ radically from country to country, and sometimes even within countries and between different water uses. It is hard to draw from the gamut of individual cases to provide a general theory, other than to say that for much modern market-incentive systems to work there is the need for a much clearer demarcation of property rights to water use that often exists. At any level, the definition of property rights to water is very difficult in the face of water's sacral quality in many societies, its essentiality to life, and its pervasive externalities. As a result, a large body of law has grown up around water in almost every country.

Water pricing, therefore, is not a task to be left solely to economists. The political, legal, and social dimensions are extremely important and emphasize the need for a "political economy" of water pricing.

3.2.13 Marginal Cost Pricing

Three important concepts tend to get confused in the discussion of water pricing. These are, (i) the opportunity cost of water discussed at length above, (ii) marginal cost pricing, and (iii) cost recovery. Marginal cost pricing is the pricing of water at the cost of supplying an additional unit of water. In a perfectly functioning market economy the most efficient solution occurs when commodities are priced at their marginal cost. The market is said to clear when the marginal cost is equal to the marginal benefit. The marginal cost must include the opportunity cost of the water. Cost recovery refers to pricing of water to recover the costs of providing it. Typically, cost recovery is based upon accounting procedures that are based upon historical costs.

3.2.14 Cost Recovery Pricing

Cost recovery pricing can be close to marginal cost pricing when the costs of new projects are like the costs of past projects. Unfortunately, in water resources development historical costs are typically much lower than current or projected costs of projects. Many water planners confuse cost recovery with marginal cost pricing, the economist's "golden rule". Most water utility officials consider marginal cost pricing as leading to unrealistically high tariffs. Unfortunately, economists often try to mandate strict marginal cost pricing, ignoring practical problems faced by utility managers. In fact, there are many different tariff structures that would allow full cost recovery without marginal cost pricing.

Brown and Sibley (1986) showed for the telecommunications industry that the Ramsey prices necessary to cover costs and profit were only in the range of 3 per cent more efficient than tariffs based upon traditional cost recovery methods. (Remember that telecommunications industry, however, has declining costs not increasing costs like the water industry.) While tariffs have financial, economic, and political dimensions, it should be remembered that marginal cost pricing relies only on the economic dimension.

3.2.15 Regulation and Efficient Pricing

The best approach probably lies somewhere between the two extremes. The tension between regulation and efficient pricing will always remain a source of contention between the planners and the political process, however, as Brown and Sibley (1986) stated:

> When regulation violates efficiency criteria seriously enough, not only economists become concerned. Peak load pricing in electricity, for example, was taken seriously in the U.S. when it became clear that excessive use of electricity was contributing to an energy crisis that alarmed many people. Hence, normative economics plays an important, though not preeminent, role in the regulatory process.

Pricing is the major tool by which economists attempt to bring about efficient resource use. Pricing is more than cost recovery; it also aims at leading society to correct allocation decisions. Unfortunately, when many of the prices in an economy are artificially set, it is hard to induce rational resource allocation by "correctly" pricing just a few of the resources. This is a major problem in all countries and regions. For example, it is estimated that in the EMENA region alone the subsidy to agriculture due to incorrect water pricing is of the order of $40 billion per year.

The Bank must develop strategies that will enable countries to bring their resource pricing progressively closer to market pricing, economy wide, than is currently the case. The Bank's Policy Paper will have to lay out the issues and consequences of price policy so that price policy will become an integral part of the Bank's own operations as well as of those of the borrowing countries.

3.2.16 Cost Recovery

In the Indian state of Bihar, the charges for Irrigation Department water from river diversions and dams are so low and the attempts to collect the fees so feeble, that in 1979 the cost incurred in collecting the fees for water was 117 per cent of the actual amount collected. Owing to its serious financial difficulties, the Irrigation Department has had to defer maintenance, and the water system has deteriorated. Farmers have less incentive to pay for poor service, and a vicious cycle has developed. Moreover, the low water charges have led farmers to use water inefficiently. By contrast, in both the Punjab region in India and in areas of the United States that rely on water from private wells, farmers are far more conservative in how much water they use since they pay higher prices. (Well water costs more because it is not subsidized.) Yet these farmers' crops are not doomed to economic failure: when the cost of pumping in the Punjab increased in the 1960s, farmers began growing more water-efficient crops such as high yielding wheat, and cash crops such as cotton, tobacco, and oilseeds.

3.2.17 Cost Recovery Rules for Bank Loans in Different Sectors

Outside observers are surprised by the disparities among cost recovery rules for Bank loans in different sectors. In the electric energy and telecommunications sectors the Bank insists on full cost recovery for the capital as well as O&M expenses. The goal is to create viable public utilities, able to set tariffs that will efficiently finance the future development of the sector. In contrast, water sector projects typically require only repayment of the O&M and of some "reasonable" part of the capital costs.

Many would argue that this policy leads to sloppy performance on the part of both the suppliers and consumers of water. Since we do know that water use is directly related to the prices charged; the higher the prices, the lower the usage.

The Bank should carefully study the implications of moving toward full cost recovery for water sector investments, and alternative methods of doing so.

3.2.18 Environmental Impacts of Water Development

There are increasing pressures on international institutions to take the lead in responding to concerns about environmental deterioration. The Bank should take the initiative in dealing with these problems, instead of its being merely reacting to specific challenges as they arise. As the industrialized nations have discovered, assuring water quality entails more than the regulation of industrial and municipal wastewater discharges. Nonpoint sources such as agricultural run-off are now recognized as major contributors to environmental problems that cannot be ignored. More importantly, at the level of questions of sustainability of ecosystems, water use and water contamination play major roles. They cannot be easily separated into water quality and water quantity issues; both are critical.

3.2.19 Casual Way of Handling Environmental Consequences in the Last Minute

From the environmental point of view much of the concern with water projects stems from the casual way in which environmental consequences are handled as a last-minute consideration, rather being integrated into project design from the beginning. Big efforts include irrigation projects in which drainage issues are neglected in the initial planning, navigation projects in which the disposal of dredged materials is haphazard, and flood control projects which ignore flood-plain fisheries. In these cases, adequate pre-planning guidelines could substantially reduce the environmental impacts. There are other projects, for example large dams, which have major environmental impacts that cannot be mitigated, such as loss of habitat, loss of fertile bottom lands, destruction of forests, and sediment trapping, in addition to the social impacts of moving large numbers of people from their homes and farms. These consequences can only be addressed on a political level by providing the right kinds of institutions that will allow for discussion, negation and compensation to the affected groups. However, guidelines that force a thorough search for alternatives could, in the case of large dams, suggest attractive alternate solutions such as modifications of the scale of the project or other storage options such as underground storage or a larger number of small storages closer to the users. The World Bank Operational Memorandum 4.00, Annexes A and B, could provide a good starting point for the development of appropriate guidelines.

3.2.20 Social Impacts of Water Development

In a perfect world, water development projects would have only positive impacts on societies. However, because of the externalities inherent in many water investment projects, there are almost always some winners and some losers. Local people lose their lands so that urban populations can have electric power and lowland fanners can have irrigation. Upstream users pollute rivers with wastes or choke them with sediment, causing severe damages to downstream users. Attention must be paid in planning projects to minimize disruption outside of the project area and to provide compensation to the affected parties.

As in the case of environmental consequences, many negative social impacts could be avoided by careful preplanning. Some of the most egregious social impacts follow from involuntary movement of populations out of areas that are flooded by dams and embankments. The move itself may be unavoidable, but social disintegration and destitution are not, and the needs of the affected populations should be integrated early into the planning process, with adequate resources and communication channels provided to meet those needs. Other cases, such as equity in water access in irrigation projects, are less obvious and are therefore often ignored by project developers and funding agencies. Impacts on women and children are also often overlooked until it is too late to mitigate their damage. Human health issues are often neglected in evaluating water sector investments. For example, the introduction of schistosomiasis into a region by irrigation projects is a serious matter and can be avoided or controlled to a significant extent by changes in system design and operation. Access to water for bathing, for religious purposes, for watering animals and for recreations are all important social uses of water often inadvertently left out of project plans.

3.2.21 Water Investments to Achieve Poverty Reduction or other Equity Goals

There is also the option of actively using water investments to achieve poverty reduction or other equity goals. The irrigation literature is replete with discussions of how to define equity, how to measure it and its effects on productivity and other values, and how to plan for it. Some options are more equitable than others and should be pursued in the design and implementation of water resources works. Nevertheless, if one were looking for investments in the economy to achieve equitable distributions of benefits, water projects are not the most likely candidates. Even within irrigation projects themselves it is important, for example, that the economic costs of skewing water allocations in favour of equity that are imposed upon society be defined in terms of the number of farmers, rather than farm sizes, and that they be carefully estimated and politically assessed as to whether they exceed the accrued social benefits. In some situations, farms are too small to be economically efficient and the water might be better used on slightly larger farms.

Equity is an emotionally charged issue which the Bank can help borrowing governments weigh properly by providing a dispassionate discussion of it in its Policy Paper. There is a large role for the Bank in helping countries deal with these types of impacts in a straightforward way by giving them due weight in their own indigenous planning methods and organizations. Some of these resources are already covered by the Bank's own environmental assessment guidelines. They should be made explicit in the Policy Paper.

3.2.22 Conservation and Recycling of Water

One of the most obvious ways of extending the water resources base is by conserving water or by recycling it after use. Unfortunately, by themselves these measures are not the panaceas that they are often thought to be, in most cases people and industries "waste" water only because it is the cheapest thing to do. Unless water prices are raised significantly there is no incentive to do otherwise. The experience in the industrialized countries is that people can be motivated to conserve water based on altruism only in times of shortage. Typically, after the stress is removed consumption rises to the previous levels.

3.2.23 Recycling of Sewage Water for Usage of Agriculture

There is much discussion of the recycling of sewage for use in agriculture. Many cities have been doing this successfully for some time. The practice is widespread in Asia, with, however, some problems in protecting the public from pathogens. With care such use can be made of sewage, and should be made wherever appropriate. However, much of the discussion implies that large amounts of irrigation can be accomplished with such recycled water. One needs to be reminded of the tremendous asymmetry in volume between agricultural and domestic water uses: one large irrigated farm in the U.S. consumes as much water as a town of 15,000 people produces wastewater. It takes the wastewater of 50 people to irrigate the land required for growing the food for one person. Although water recycling for industrial and domestic uses appears attractive and it is a worthwhile investment for vegetable gardening, it is not likely to make a major contribution to irrigating field crops.

3.2.24 Recycling of Municipal Waste Water Without the use of Irrigation as an Intermediary

Direct recycling of Municipal wastewater has also been practiced without the use of irrigation as an intermediary. For example, Windhoek, Namibia, has been successfully recycling large quantities of municipal water for more than 20 years without apparent problems. In some parts of the world such recycling is practiced in an unobtrusive way. For example, in Holland polluted water from the Rhine is filtered through the river banks before being abstracted for urban water supplies.

A lot has also been written about conserving water by improving the efficiency of use in irrigation and reducing the losses in urban water systems. As discussed below, increasing water use efficiencies in agriculture is a difficult concept and the practice may not really provide great savings of water in many cases, particularly where conjunctive use of groundwater is practiced. Nor is it necessarily wise to reduce losses in urban water systems, since the cost of reducing the losses may exceed the benefits as perceived by the water utility. This water is clearly an area where the appropriate pricing and valuation of the water itself is the major issue. Conservation for its own sake is not a realistic or advisable goal; it only makes sense within a correct pricing policy for water. The Bank's Policy Paper should clarify the potential role of conservation and recycling.

3.2.25 Host Country Institutional Reforms

In most countries, there is a great need to upgrade the institutions dealing with the various aspects of water. Most importantly, in every country there should be one institution with the responsibility to coordinate water policy across uses and across government agencies. Even advanced industrial countries such as the United States, suffer from lack of coherence in water policy across uses. There is often a tremendous inertial asymmetry between the staffing of agencies with traditional concerns and those with more modern ones. For instance, in Thailand the Irrigation Department is the single largest government agency. In India, the use of surface water is administered by the Central Water Commission, whereas the Central Groundwater Board oversees groundwater use. There is little practical integration of these agencies' plans, despite a formal agreement to cooperate. Each of India's 22 states also has its own department of irrigation, which usually deals only with surface-water supplies, while other departments focus on groundwater. In cities around the world, municipal water is a local responsibility, but many individuals also sell water in small quantities to households.

Rural households are almost always in charge of obtaining their own supplies. The result of such compartmentalization is that agriculture, industry, and municipalities use water inefficiently.

While institutions are likely to be highly regional and culturally specific, the need to strengthen them is one of the major generic problems facing development activities around the world. The need is not for large, new institutions but, rather, for small policy councils composed of cabinet rank officials who will be able to coordinate the work of the existing water institutions. The Bank's Policy Paper should emphasize structure, organization, management, and the role of adequate cost recovery, to enable individual countries' institutions to carry out necessary maintenance tasks and build for future demands.

3.2.26 Role of the Bank in Education, Training, and Information Exchange

As the world's leading institution for supporting investments in the water sector, the Bank is ideally positioned to use its prestige and influence to further education, training, and information exchange. Through its Economic Development Institute (EDI) the Bank is already heavily engaged in the formal training of Third World government servants. This role could be further enlarged by requiring more on-the-job training for host government employees. As mentioned above, the appraisal process itself could be used as a useful teaching mechanism. The environmental and social impact assessment requirements stressed above can be used as a public education tool. If government agencies develop and use environmental and social impact assessments, they will discover, as will the impacted groups if they are integrated into the process, how water investments really function well in their own societies. This has been one of the most important outcomes in those developed countries which have implemented such an approach. Their citizens are now much more aware of the environmental and social consequences of government action and have begun to make their views felt through political action. Green parties are springing up throughout the industrialized countries. Since funds are limited, the Bank might consider a role for itself in educating other public and private financing agencies in this field.

3.3 Subsector Issues

3.3.1 Irrigation

"Water and food" carries many assumptions about development policy which need to be carefully and separately examined. Food security is a major concern of every government throughout the world but how to achieve a satisfactory level of security is not obvious. National self-sufficiency in food grains is not the imperative as it is often assumed to be. In the modern world oil is essential for survival, but nowhere do we find the belief that every country should be self-sufficient in oil. There is no reason why every country should grow grain, which requires 2,000 to 3,000 tons of water per ton. In an arid country water can have much higher value for other uses. International trade can redistribute water – in the form of grain – to nations that decide not to grow it. Israel subscribes to this practice, emphasizing valuable cash crops such as fruits, vegetables, and flowers for the European market. Countries that take this approach could maintain stockpiles of the basic grains; even though stockpiles are costly and can deteriorate; replacing spoilage is less expensive than growing grain. Such countries could grow a diversity of cash crops so that fluctuating international prices could not devastate their economies.

Even the definition of self-sufficiency is not clear. Does it mean that a country should be self-sufficient on the average, or for drought years, or only 75 per cent of the time? The choice of the definition has a large impact on planning for water use in agriculture. It is imperative that the Bank address this issue head-on in its Policy Paper.

3.3.2 Increasing "Water Use Efficiencies" in Agriculture

Increasing "water use efficiencies" in agriculture is a widely-suggested solution to water shortages. However, in many parts of the world one farmer's inefficient use provides the water for another farmer downstream or recharges the groundwater for many other farmers. It has been argued that the irrigation efficiency in California's San Joaquin Basin approaches 100 per cent even though on any one farm the efficiency may only be 60–70 per cent.

Individual cases must be closely considered before much can be made of the advisability of improving "irrigation efficiency". Many commentators looking only at farm level efficiencies have advocated improvement of efficiency of water application as the best way to conserve water. Although this may be true in some cases, it is not universally true and should be subjected to careful analysis.

3.3.3 Economic Efficiency of All Water Use Rather than of "Irrigation Efficiency"

It is more appropriate to start from the larger perspective of the overall regional efficiency of the use of water for irrigation. But even this is a difficult concept, and agricultural and irrigation professionals will be on still stronger ground if they orient themselves to the further-reaching concept of the "economic efficiency" of all water use rather than of "irrigation efficiency". "Irrigation efficiency" is a seductive notion because it simplifies the problem to water alone by focussing it on engineering and management variables. But it can become a form of tunnel vision; it can throw the larger benefit-cost picture seriously out of balance, and lead to misplaced investment on a significant scale.

One will have to devote a major effort to analyzing irrigation since this is one of its major lending activities. However, there are many other issues involving choice of technology and cropping policies that have not been featured here, which are also of major concern in this part of the water sector. Many of the sector-wide issues apply in the irrigation area. For example, a recent paper by Johnson (1990) indicates that the problems with irrigation in South and Southeast Asia are becoming particularly acute because of the funding of the operations and maintenance (O&M) of the systems. As indicated above, pricing and cost-recovery are fundamental to achieve proper operation and maintenance of systems.

3.3.4 Water and Health

A relationship between water and health has been accepted since long, the Water Commissioner of Rome in 97 A.D., but the exact relationship is not well understood even now. A World Bank position paper on domestic water supply (World Bank, 1976) cautiously limited itself to saying that "other things being equal, a safe and adequate water supply are generally associated with a healthier population". Many of the benefits observed during the "sanitary revolution" in the industrialized world were achieved when protected water supplies were provided to societies with rising income levels, which induced other positive behavious with respect to personal hygiene and nutrition. At low levels of development, investments in improved domestic water supplies are often necessary but generally not sufficient to realize the potential health benefits. For middle-level developing countries, where the population is generally better educated, the health benefits of investments in water supply and sanitation are generally substantial. At higher levels of development one typically observes small additional health benefits from further investments in such facilities.

3.3.5 Area Developed through Water Resources Introduced Serious Human Health Problems

But apart from drinking water and sanitation, serious human health problems can be introduced into an area through the development of water resources. The classic case is the spread of schistosomiasis in Africa and Asia after the introduction of irrigation. Debilitating rather than fatal in most cases, schistosomiasis now infects more than 200 million people. Malaria, filariasis, yellow fever, onchocerciasis (river blindness), dracunculiasis (Guinea worm disease), and sleeping sickness can be spread by irrigation projects. Many water-soluble chemicals used in agriculture, particularly chlorinated hydrocarbons, accumulate in animal and plant tissues and thus enter or become concentrated in the food chain. High concentrations of nitrogen fertilizer in water supplies cause blood diseases in infants. In many countries unregulated and indiscriminate production and disposal of chemicals and their wastes have led to serious contamination of surface and ground waters by toxic and carcinogenic substances. Some of the worst cases are observed in developing countries which have few resources to deal with them. A careful review of these components of "water and health" is in order, not just of drinking water and sanitation.

3.3.6 Urban Water Supply and Wastewater Disposal

With continuing rapid population growth, the number of cities of over one million people is expected to increase rapidly. It is in these burgeoning metropolises of the Third World that the most pressing needs for water supply and wastewater disposal investments will be felt. In all cities, the costs of providing new water supplies and wastewater facilities are rapidly increasing. Careful assessment of new approaches to the whole problem is needed. For example, Akhter Hameed Khan, the former head of the Comilla Academy for Rural Development, has successfully applied the approaches of rural organization to the densely-packed parts of the city of Karachi, mobilizing the local population to plan, construct, and finance a sewer and drainage system at a fraction of the cost of similar systems provided by the city government.

3.3.7 Innovative Ways of Dealing with Urban Water and Sewer Issues

We must look harder for innovative ways of dealing with urban water and sewer issues. Technical and managerial performance needs scrutiny to ensure that excessively costly systems are not built merely to copy the industrialized countries. This is particularly true now in Eastern Europe, where the tendency is to proceed immediately with the same approaches as those used in Western Europe. As is well known, the general rule is that removing the first 50 per cent of pollution is quite cheap, with the incremental improvements being increasingly costly. This consideration often supports developing a system in stages, moving to the expensive higher levels of performance as general economic growth generates increased ability to pay for them.

3.3.8 Alternative Approaches to Urban Water Supply and Waste Disposal

The costs of alternative approaches to urban water supply and waste disposal should be carefully examined. Local agencies should be made aware of these costs and encouraged to set their tariffs to reflect real costs and projected trends of costs. The issue of cost recovery, discussed in more detail above, is critical here. The Bank should also initiate studies to elucidate the economic benefits of the provision of these urban services. Such studies should be based upon assessments of the willingness-to-pay for these services by different socioeconomic groups. Such studies are conceptually difficult as well as time-consuming, but they should be carried out in a sufficiently large number of cases that the basic magnitudes of the expected benefits can be demonstrated. This is particularly important in this sector since most of the literature focusses upon the cost of the services and on complaints that they are becoming too expensive. "Too expensive" compared to what? The only meaningful comparison is with the benefits received from consuming these goods and services.

3.3.9 Role of Industry in the Management of the Ambient Quality of the Water Resources

For urban water and wastewater disposal, consideration of the role of industry in the management of the ambient quality of the water resources is vital. Industry uses only 5 to 10 per cent of all water supplied but still represents an important segment of demand. This results in part from the fact that industrial processes pollute a disproportionate amount of water. For example, in Sao Paulo and Seoul, industrial pollution has turned many streams and rivers into open sewers. Developing countries should learn from the experience of industrialized nations: it is much more expensive to clean up polluted water than it is to avoid polluting in the first place. Economic incentives against industrial pollution, such as effluent fees, should be established so that companies will have an incentive to control their effluents at the source. In addition to the usual regulatory methods, innovative approaches such as tradable permits and privatization of facilities should be explored. The failure of some of these approaches in the industrialized world does not mean that they may not be appropriate in Third World settings.

3.3.10 Regulating Environmental Quality by Pricing

Regulating environmental quality by pricing the effluents that individuals, municipalities, and corporations emit is one of the standard recommendations of economic theory. The economic literature is full of conceptual schemes for pricing and taxing effluents that would lead to internalizing the externalities of pollution. Once this is done, environmental quality can be left entirely to the usual market forces. The Washington-based research institute.

Resources for the future, has led the campaign for fees on effluent discharge into waterways for over 20 years with little success; it has mainly been opposed by those who think it improper to sell the right to pollute the environment.

3.3.11 A Substantial Amount of Effluent Pricing in the U.S. and in Europe

Nevertheless, a substantial amount of effluent pricing has been used in the U.S. and in Europe. Hudson (1981) reports on the extent and mode of implementation of water pollution pricing through sewer charges. He claims that by 1970 more than 90 per cent of the municipalities with populations over 50,000 levied some form of sewer charge on residences and industry and that 40 per cent of local expenditure on sewage were derived from these charges. Industry can choose to pretreat its waste, decrease its water use, improve housekeeping, change either the production process or products, or it can choose to pay the effluent fee. The industrial charges are typically related to die quantity of water used and the "strength" of the effluent measured in terms of the oxygen-demanding organic waste load (BOD) and total suspended solids. These charges can give an incentive to industries to change the amounts and strength of their sewage effluents. Hudson's study of five large cities (Atlanta, Chicago, Dallas, Salem (Oregon), and South San Francisco) and 101 industries found that effluent charges were overwhelmingly preferred by the industrialists to discharge limitations. There was a universal attempt by the industries in the sample to respond to the effluent fees despite the irrelatively small costs to the industries. Hudson concluded that; "...we are confident that economic incentives work well and can be effectively administered". In its Policy Paper the Bank should stress the application of well-established economic principles and innovative technical approaches in this area.

3.3.12 Rural Water Supply and Sanitation

This one problem area could easily consume the entire Bank lending portfolio and still not be completely resolved 1990 is the end of the United Nations Water and Sanitation Decade, but in percentage terms the population covered remains like that at the beginning of the Decade. More effort is needed on the social and economic side of this issue, and Bank sponsored research on assessing willingness-to-pay for rural water supply should be extended to sanitation. In addition, there is a need for research on appropriate incentive schemes to encourage individual or private sector undertakings to provide goods and services in rural water supply and sanitation.

The Bank has played a major role in the Water and Sanitation Decade and it would be appropriate to include in its Policy Paper a critical review of the accomplishments and problems with respect to this entire set of activities.

3.3.13 Hydropower

Hydropower is a major area of Bank funding and one that until recently was considered well defined and not contentious. Recently, however, hydro projects have received a lot of criticism based upon their environmental impacts. The Bank's environmental guidelines have already begun to address these issues and should make sure that governments receiving the funds are equally as careful in the development of hydro projects. Some economic issues that need clarification in the hydro area relate to the inter sectoral nature of electricity and the conflicts over the potential uses of the water between users.

In addition, the pricing of the output from multipurpose water resources projects that include hydropower needs careful reappraisal.

3.4 Conclusion

The temptation is to load a disproportionate amount of the cost recovery on the electricity consumers and not on the agricultural and other users of the joint output of the project. The above discussions have helped to develop explicit guidelines for dealing with these issues, particularly the environmental and social impacts, in its own operations and to encourage the governments receiving funds to recognize them as potential problems.

References

Baumol, W.J. and Oates, W.E. *Economics, Environmental Policy, and the Quality of Life.* Englewood Cliffs: Prentice-Hall, 1979.

Brown, S.J. and Sibley, D.S. *The Theory of Public Utility Pricing.* Cambridge: Cambridge University Press, 1986.

Dasgupta, P., Marglin, S.A., and Sen, A.K. *Guidelines for Project Evaluation.* New York: United Nations Industrial Development Organization, 1972.

Desvouges, W. H. and Smith, V. Kerry. Benefit-Cost Assessment for Water Programs, Volume I. Prepared for the U.S. Environmental Protection Agency. Research Triangle Institute, North Carolina, 1983.

Dorfman, R., and Dorfman, N. *Economics of the Environment.* New York: W.W. Norton, 1972.

Coase, R. "The Problem of Social Cost." *Journal of Law and Economics,* Vol. 3, October, 1960, pp.1–44.

Eckstein, O. *Water Resource Development: The Economics of Project Evaluation.* Cambridge: Harvard University Press, 1958.

Falkenburg, M., Gam, M. and Cestti, R. "Water Resources: A Call for New Ways of Thinking". World Bank, INUWS Working Paper, March, 1990.

Fisher, A. and Raucher, R. "Intrinsic Benefits of Improved Water Quality," in V. Kerry Smith, (ed.), *Advances in Applied Micro-Economics,* Vol 3. Greenwich Connecticut JAI Press Inc., 1984.

Hanke, S.H. "Economic Aspects of Urban Water Supply: Some Reflections on Water Conservation Possibilities". International Institute for Applied Systems Analysis, Laxenburg, Austria, December, 1982. Mimeo.

Hanke, S.H. and Davis, R. "Potential for Marginal Cost Pricing in Water Resources Management," *Water Resources Research,* Vol. 9, No. 4, August, 1973.

Hirschleifer, J., Milliman, J.W. and De Haven, J.C. *Water Supply: Economics and Policy.* Chicago: University of Chicago Press, 1960.

Hudson, J.F. *Pollution Pricing: Industrial Responses to Wastewater Charges.* Lexington, Mass.: Lexington Books, 1981.

Hufschmidt, M.M. and Dixon, J.A. Economic Valuation Techniques for the Environment: A Case Study Workbook. Baltimore: The Johns Hopkins University Press, 1986.

CHAPTER 4

Market Failure in Water Resources Planning

4.0 Introduction

The literature since Eckstein shows how each of these market failures can be compensated for in water resources planning. It is now the general opinion that careful application of neo-classical principles can deal with the market failures most likely to arise in water investment analysis. For example, increasing returns to-scale and monopoly pricing can be dealt with by some form of Ramsey pricing. Externalities can be dealt with by expanding the definition of the system to "internalize the externalities". The lack of mobility of resources can be accounted for by suitable shadow pricing, and the income maldistribution can be dealt with by adding constraints on the distribution of benefits. The adjustments themselves lead to some loss in the overall "economic efficiency" implied by the classical model, but the loss is believed to be small relative to the broader improvement of solutions. Brown and Sibley (1986) discuss various techniques of improving utility pricing which come close to the economically efficient solution and which, at the same time, are closer to the practical methods useable by utilities.

4.1 Valuation of Water

Allocation of water among the myriad conflicting uses presents a major task to governments, all of which take responsibility in some degree for regulating access to water. It is difficult to assign unambiguous economic values to many uses, and hence these may be implicitly overvalued, undervalued or completely ignored in the decision-making process. Rogers (1986) gives many examples of the problems that arise from undervaluing water.

Many of the problems of valuing water stem from the market failures mentioned above. The existence of externalities and the lack of mobility of resources make finding the market price quite difficult. In a perfectly functioning economy envisaged by the classical economic model is that "price equals value", and the cost of providing a good, after allowing for payments to all its factors of production, will precisely equal its market price. Because of this elegant solution one only must establish "cost" to establish "value".

4.2 Equating Water Costs with Its Value

Unfortunately, many water resources planners forget that simply equating cost with value only holds true in a perfectly functioning market economy. In all other cases (that is almost all cases) care must be taken not to confuse cost with value. What then is the "value" of water? The answer appears to depend upon "to whom" and for "which use". Drinking water is obviously valuable and becomes increasingly so as the amount available decreases. A glass of water could be infinitely valuable to a person dying of thirst in the desert but not very valuable to a woman living alone on the banks of a pristine river. In the second case the woman would only be willing to pay the cost of somebody going to the river and fetching the water for her. She would be unwilling to pay more because she could go and take it herself. So, the value in this case is the cost of obtaining the water. Now, if there was a farmer irrigating land alongside the river, how much would the water be worth to him? If there is enough water in the river so that the woman can have as much as she can drink just at the cost of obtaining the water from the river, then obviously, the farmer can take as much as he wants at the cost of obtaining the water.

Clearly, the farmer would also value the water at the cost of obtaining it. So far, so good; cost equals value.

4.3 Typical Users of Water Resources

However, such bucolic settings no longer exist in the modern world. Typically, there are many users of the resource apart from the housewife and the farmer. At some point in time the use by one person will start to interfere with the use by another. At that point, the water is said to have an "opportunity cost". Since the continued abstraction by one user reduces the amount available to another there is a loss of the opportunity to use the water by one user. This lost opportunity costs the affected user the amount he values these units of water. At this point the "value" of the water should reflect the willingness-to-pay of the user who is losing water. If for some institutional reason the housewife must cut her consumption of drinking water, then the opportunity cost to society of this allocation of water away from her is her willingness-to-pay for water. If the allocation of the water shortage were the other way around, the relevant opportunity cost would be the farmer's willingness-to-pay for irrigation water.

If the question of how to allocate the water were left to an outside party, for instance, the World Bank, then that party might ask how society would best benefit from the allocation. One way of answering this question is to apply the logic of social choice theory embodied in modern economics, which allocates the water to the use with the highest value.

4.4 Estimation of Opportunity Cost of Water

Establishing the willingness-to-pay various consumers of water is fairly a well-developed field in economics and can be easily adapted in many water conflict situations to establish estimates of the opportunity cost of water. Unfortunately, many economic studies of water use ignore the opportunity cost of water and only reflect the actual costs of obtaining "the water itself". As mentioned above, if there were well-established markets for water then the market price would itself reflect the opportunity cost of water. However, in most countries such markets do not exist and one is left to estimate the opportunity cost in indirect ways.

4.5 Opportunity Cost of Water Zero When There Is No-Shortage of Water

The opportunity cost of water is only zero when there is no shortage of water. In evaluating water investments, the value of water to a user is the cost of obtaining the water plus the opportunity cost. Ignoring the opportunity cost part of value will undervalue water, lead to failures to invest, and cause serious misallocations of the resource between users. The opportunity cost concept also applies to issues of water and environmental quality.

4.6 Planning for Water Resources

The management imperatives of water are embedded in its nature as a scarce resource that is often treated as a public good. External effects occasioned by water use also invoke important economic imperatives. Property rights and externalities raise political and financial imperatives which influence the choice of ways to price water to pay for the public supply of it. To appreciate fully water policy options and how they are evaluated, it is necessary to understand how economics is used and misused in the water area. Even though politics ultimately controls water resource planning, the discussion is usually framed in economic terms, and the ability to understand and manipulate the economic analysis may significantly improve the outcome. Hence, a significant part explores the possibility of improving the economic aspects of planning. The attention devoted to economics should not be taken to mean that the institutional and technological dimensions are unimportant, but that in the judgement of the author the pay-offs from improving the economic dimensions are currently larger than those from other areas of concern.

4.7 Current Planning Approaches – Analysis of Benefits, Costs, and Technology Choice

Water resources planning has received a large amount of attention from economists and planning professionals for many decades. Based on the Flood Control Act of 1936, federal agencies in the United States, for example, were charged with devising economic criteria to ensure that only water projects for which the "benefits exceeded the costs" would be implemented by the federal government. This policy led to major research that culminated in the 1960s in fundamental works by Eckstein (1958) and Maass et al. (1962) at Harvard University, Hirshleifer, Milliman and DeHaven (1960) at the University of Chicago; and Kneese and Bower (1968) at Resources for the Future in Washington, D.C. The finishing touches with regard to dealing with environmental quality were in place by the end of the 1970s, with works by Dorfman and Dorfman (1972) and Baumol and Oates (1979). In addition to these books there are literally hundreds of other excellent texts explaining many of the practical aspects of the detailed analysis of benefits, costs, and technology choice for each possible use of water. While this was taking place in the water resources field, the Bank, UNDP, and many other funding agencies were developing reliable methods for economic appraisal of investment decisions in general. Outstanding examples of this literature are Little and Mirrlees (1974) and Squire and van der Tak (1975), sponsored by the World Bank, and Das Gupta, Marglin, and Sen (1972) and the Guide to Practical Project Appraisal (1978), sponsored by the United Nations Industrial Development Organization (UNIDO). These works are the basis for the current evaluation of investments in the water sector.

4.8 Planning Water Resources – River Basin and Individual Project Levels

The literature provides sound methods for planning water resources at a river basin level and at the level of individual projects. However, since most of the methodologies were developed in western industrialized nations, they pay little attention to planning for water resources in a macro or inter-sectoral way.

State-of-the-art economic project analyses are carried out routinely by the Bank staff. For example, benefit cost analysis is carried out on irrigation assessments and the project is recommended only when a suitable internal rate of return (IRR), or another index such as a benefit-cost ratio is achieved. Recently the choices have also been conditioned upon not causing "too much" social or environmental harm. For projects such as urban water supply or sewerage, where benefit calculation is very difficult, the marginal costs of providing the services are computed and compared to other ways of providing the same services. For both types of project analysis, care is taken to correctly shadow-price the inputs and outputs. However, to decide whether to spend money on irrigation instead of on flood control, or on other investments outside the water sector, the rules for project analysis noted above need to be bolstered by additional assumptions and theory. In the theory, pricing of the resource is critical to the correct appraisal of the project and the correct implementation.

4.9 Pricing Has Three-Fold Role in Water Policy

Pricing has a threefold role in water policy. First, the price can be set to recover the cost of the investment from the beneficiaries. Second, increasing prices tends to ration water by cutting" uneconomical consumption. Price increases cut the demand for water by moving up the demand curve, which is the effect most decision makers look for when pricing policy is advocated. The third aspect of pricing, and the one most frequently overlooked, is that of increasing the supply. When the price is higher, supplies of water from more expensive sources become available.

4.10 Price an Integral Part of the Definition of "Scarcity"

Price is also an integral part of the definition of "scarcity". The issues of pricing and cost recovery are a fundamental part of any approach to deal with the modern realities of water planning. The fact that pricing has not been rigorously applied in the past, and that the current prices of water are still generally quite low, is an indication that scarcity is a relatively new phenomenon. It also gives confidence that, at least for the next decade, moderate price increases will solve the water scarcity problem. Pricing and cost recovery are an integral part of the resolution to many of the water issues discussed below.

4.11 Where Does the Current Approach Break Down?

The Bank's Operations Evaluation Department has shown that current planning approaches worked well when the competition for water was less acute. The same methods should also work under the new conditions. That they are not working is caused by a lack of attention to the correct implementation of the methods. For example, even if correct prices are computed in the appraisals, they are not used by the borrowing governments in implementing the projects once funded. This fact leads to overconsumption of water by some users and artificial shortages for others. There are two major areas where the current approach seems to be in trouble:

4.11.1 Using the Concept of Opportunity Cost of Water

Establishing and using the concept of opportunity cost of water in different sectors of water use. Since the methods for evaluation are applied mainly to project-by-project appraisal, consideration of the inter-sectoral nature of water use is neglected. Therefore, perfectly well analyzed Bank irrigation projects in Algeria, for example, find themselves in direct conflict for the same water with other, equally well prepared Bank projects for urban water supply. In such cases the relative marginal benefits of additional investments in water must be carefully compared with those of other sectoral investments. Other resource sectors, such as energy, have well developed methodologies to relate sectoral and macro plans. Most of these were developed in response to the oil crisis of the early 1970s. It is now possible for energy planners to show the macroeconomic consequences of regulation of, and investments in, various energy sources, and, in the other direction, to estimate the consequences of shifts in macro policy on demands and supplies for energy by various sectors. This has not been done in the water sector.

4.11.2 Water Investments as Infrastructure Services

The role of water investments as infrastructure services which serve as both intermediate goods and final goods is also often overlooked in water planning. So, for instance, Ingram (1989) expresses concern that water infrastructure may suffer under-investment because of the bank's structural adjustment programmes for various countries.

A fundamental piece of information missing in most water plans is the opportunity cost of water. Adding the opportunity cost of water to its marginal cost of supply and comparing this with its current price indicate show efficient current water use is and how efficient it could be. To estimate these numbers, it is necessary to have some form of inter-sectoral comparison of value or willingness-to-pay, hence the need for a strong policy directive in the Bank's Policy Paper to establish methods to evaluate water investments in an inter-sectoral context.

4.11.3 Social and Environmental Concerns into Planning

Incorporating social and environmental concerns directly into planning, social and environmental impacts of Bank projects are increasingly a source of contention between the Bank and environmental interest groups in developed countries. It would be possible for the Bank to undertake economic analyses of the environmental impacts of water projects more diligently than has been done in the past. There is no doubt that this is a difficult task; nevertheless, there is now a large literature dealing with the economic impacts as lower bounds on the total impacts of a project.

> A society that allows waste dischargers to neglect off-site costs of waste disposal will not
> only devote too few resources to the treatment of wastes but will also produce too much
> waste in view of the damages it causes. (Knees and Bower 1965)

As use increases, the public goods nature of water and the pervasiveness of externalities in water use lead inevitably to an increase of pollution. In the late 1960s and 1970s the public perception in the developed countries was that the environment was being "over-polluted". The externalities implied in water pollution were first extensively discussed by the Resources for the Future group in the early 1960s. The above-cited book by Kneese and Bower (1968) is still a leading text on the economics of water quality.

The concern of economists analyzing water pollution is with "market failure". The issue of how to internalize the externalities has received most of the attention. At the most obvious level if the effluent could be correctly priced, then industry and other polluters could be taxed by just this amount and problem would disappear – the correct amount of pollution would be obtained, and if this were considered too large then the effluent price could be raised until a satisfactory lower level of pollution was achieved.

4.11.4 Methodologies for Economic and Non-Economic Dimensions of Environmental Deterioration

Coase (1960) showed that externalities by themselves do not lead to economically inefficient solutions provided that the polluters and the people being affected can freely and inexpensively negotiate with each other. Coase claimed that the responsibility for damages is a reciprocal one, with the affected party taking steps to avoid them as much as the perpetrator takes steps to avoid producing them. Economic efficiency is achieved when external damages are borne by the party that can most cheaply repair them.

Hufschmidt and Dixon (1986) have extended environmental economics to the issues faced in developing countries, particularly with respect to water projects. Papers by Devousages and Smith (1983) and by Fisher and Raucher (1984) show how contingent valuation methods can be used to evaluate the benefits (and damages) associated with a wide range of environmental impacts that hitherto had been thought to be non-measurable.

Hence, there is a need to pursue this line of thinking and lay out the appropriate methodologies for assessing the economic dimensions as well as the non-economic dimensions of environmental deterioration.

4.12 How the Problems Can Be Resolved

There are two possible, and not mutually exclusive, explanations of why problems arise in the way water resources are planned, appraised, and implemented.

4.12.1 Economic Principles Adequate but Applications Inadequate

The accepted economic principles are adequate but the application is inadequate. The appraisal approaches used at the Bank and at other funding agencies are based upon the generally accepted principles outlined voluminously in the literature referred to above. The application of the principles is flawed because, when faced with economic scarcity of water its use must be priced at its opportunity cost. This is typically not done and leads to serious misallocation of resources between different uses of water. For example, in the EMENA region the opportunity cost of water in the municipal sector is at least two to three times as high as the marginal value of irrigated agricultural production per cubic metre of water for all crops except some vegetables (tomatoes and cucumber). However, when the irrigation projects are evaluated by themselves, without consideration of the opportunity cost of water in other sectors, they appear to have acceptable Internal Rates of Return (IRR).

Similarly, the externalities due to environmental damages are not integrated into the appraisal of projects. The economic principles are clear, but little attention has been devoted in the past to evaluating the damages and therefore they tend to be overlooked in the analysis.

4.12.2 The Theory Needs Refining

No theory is perfect, and so it is with the theoretical approaches to water resources management. The relevant question is how the imperfections in the theoretical underpinnings lead to unacceptable outcomes from the point of view of the Bank, other funding agencies, and governments. There are several assumptions about the theory that cause difficulty in water planning, but there are also approaches to softening or eliminating these market failures from the conceptual model. Three areas stand out as causing more trouble than others and, hence, may need more refinement of the theory.

4.12.3 Allocations between Investments in Water Resources and Other Economic Sectors

The difficulties associated with making macroeconomic allocations between investments in water resources and other economic sectors. The pervasiveness of water use throughout the economy may partially explain the first problem, but other sectors, including energy, radiate similar pervasive connections throughout the economy and society. Water planning has been an integral part of governmental planning for much longer than other resource sectors. Hence, large and powerful interest groups are fully cognizant of their stakes in planning. In most countries, it is not of concern that the water planning and investments be balanced with other sectoral activities; indeed, the idea of such balancing is threatening of groups with a vested interest in water investments and management. The development of reliable planning methodology to relate water sector plans to the overall macro development of a country is a generic problem that should guide investments and other Bank water activities.

4.12.4 Externalities Directly in the Appraisal Methodology

Dealing with externalities directly in the appraisal methodology is another important requirement. The theory and methods for doing this have been developed and many environmental consequences can be adequately treated within the existing benefit-cost methodology. These methods have only received sporadic application in actual project evaluation. Fisher and Raucher (1984) report on several such studies but additional work is needed to develop this methodology for the types of projects supported by the Bank.

4.12.5 The Political Economy of Tariff Setting in the Water Sector

The political economy of tariff setting is extremely important and must a large extent has been neglected by water experts. This is particularly true in the cases often encountered when there are economies of scale associated with the investments. The canon of price theory is marginal cost pricing. In other words, set the price at that point where the demand and the supply curves intersect. Unfortunately, establishing the marginal price for many water uses is difficult because of the characteristics of water that make its supply a natural monopoly. The existence of economies of scale ensures that those who first entered the market will always be able to underprice and drive out any potential newcomers. They can decide on their desired profit level and set the prices accordingly. This fact has been recognized for a long time in municipal water and wastewater treatment and the suppliers have been regulated, typically by public utility commissions. Meier (1983) showed that there are at least five different ways of measuring marginal cost under these conditions, each resulting in a different estimate of the marginal cost. In practice, good estimates can be obtained for most water investment projects with either of the approaches currently used in Bank appraisals: long run marginal cost (LRMC), and average incremental costs (AIQ).

Unfortunately, the most regulatory commissions have tended to Backward accounting pricing based only on average costs and revenue needs to meet them.

4.12.6 Traditional Rate-setting Methods

Traditional rate-setting methods, employed by state regulatory commissions as well as local government agencies, appear to have produced a situation of rapidly deteriorating water systems, rural and urban, characterized by aging capital facilities and under-maintained water systems. (Mann, 1981)

This stance is not appropriate in situations where these utilities are facing increasing marginal costs (in all the best projects have been built and now it becomes increasingly difficult to supply the same amounts of water at the historical costs). Under these conditions a forward-looking accounting stance is appropriate. If this policy were pursued, the emphasis on revenue requirements of utilities would be replaced by establishing adequate future investment funds.

4.13 Conclusion

Economic theory provides "second best" pricing algorithms for such circumstances. However, according to Hanker and Davis (1973) even these lead to problems for utility managers. Therefore they suggested the need to develop "third best" pricing methods, which would be more responsive to the political dimensions of water resource pricing. In a study of water conservation in Perth, Australia, Hanke (1982) outlines how to estimate economically efficient choices considering the resource cost to the utility, the resource cost to the consumers, and the useful consumption foregone because of a conservation policy. Interestingly, when he compared two pricing policies (marginal cost pricing and seasonal marginal cost pricing) with three non-price policies (leakage, control, metering and mandated water restrictions) he found that the most economically efficient was leakage control, followed by metering and annual marginal cost pricing. All the other policies had negative benefits. This is an eloquent reminder that rationing by pricing is not always the most efficient approach. To find out, however, it is imperative to carry out analysis at the level of economic sophistication applied by Hanke.

CHAPTER 5

Water Resources Management Policy: Revisited

5.0 Back Drop

Peter Rogers in his Working Papers (WPS-879), which is a joint product of the Water and Sanitation Division, Infrastructure and Urban Development Department, and the Agricultural Policies Division, Agriculture and Rural Development Department of the World Bank, defined a bank water resources management policy.

The world is facing keen competition for limited supplies of alternative water uses in agriculture, urban and industrial supplies. Further its use for recreation, wildlife, human consumption, and maintenance of environmental quality are other areas of concern. In many parts of the world manifests this competition and our current concern is deal with ability. In India, a large irrigation project could not function due to diversion of water to the rapidly growing city of Pune. In China, water shortage reduces industrial production, even though industries are surrounded by paddy fields. In California, irrigation leaches selenium salts and is killing wildlife.

Bank irrigation projects in Algeria are now competing with Bank urban water supply projects for the same water, and many proposed irrigation projects and most hydro project proposals are on hold because of environmental concerns. Until recently planners in the developing countries were adopting water planning management and the analysts at the funding agencies were, by and large, appropriate and adequate to the task at hand.

The increased competition for water has suggested that the project-by-project planning methods are inadequate. Hence, there is a need to have new approaches that will integrate water resource use across different users and across different economic sectors. The present approaches are no longer sufficient to deal with these enormously complex and difficult issues. Bank has launched a comprehensive water resource management policy study to investigate how best to resolve them. This chapter confines to the issues which should be included in the Bank's Policy Paper.

5.1 The Water Sector at the World Bank

The Water projects funded by World Bank have broad impacts on the economies of developing countries. Water is used for domestic, industrial and irrigating purposes. It is a major resource of hydro-power and thermal power generations. It is also essential to many transportation systems. Inland and estuarine fisheries depend upon adequate flows of high quality water in the rivers. Rivers, lakes, and groundwater are also extensively used for the disposal of treated or untreated wastewater, that itself has served to remove and transport chemical or other wastes.

5.2 Freshwater Stops Penetration of Salinity into Surface or Groundwater Sources of Supply

Freshwater stops the penetration of salinity into surface or groundwater sources of supply. Increasingly water access is a major component of both urban and rural recreation facilities. In many places, large investments are required to minimize the consequences of naturally occurring floods and droughts. The maintenance of a locality's fauna and flora is critically dependent upon access to water. These uses have either been funded, or are potentially fundable, by Bank resources.

The World Bank lends substantial portion in water sector. The average annual lending of the World Bank is estimated at 21 billion dollars. Thus, total lending of the Bank is arrived at 840 billion dollars. Out of a total lending of $840 billion, water projects constituted $10.6 billion, or 12.8 per cent of the total. Irrigation and drainage accounted for $4.6 billion (5.5 per cent), hydropower for $2.4 billion (2.8 per cent), and water supply and wastewater $3.5 billion (4.3 per cent). Hardly, any data is available for flood control, water transport, and river and estuarine fisheries. About 55 per cent of the total expenditures for agricultural and rural development was spent on water projects. In many countries (for example Brazil) water sector expenditures are as high as 30 per cent of total public expenditures.

5.3 The Bank and Its Central Departments Deals with Water sector independently

The Bank's regional bureaus and its central departments deals with water sector independently. This situation has sometimes created inconsistencies in Bank policy between countries or between water uses. In the event of existence of water surpluses, such inconsistencies are not necessarily bad, because of wide diversity of the countries and range of uses for water. Indeed, this is the observation of the Bank's own Operations Evaluation Department. As such the Bank has carried out post-completion audits for many projects (average 150 projects). Irrigation projects, for example, performed quite well in the arid regions in Europe, the Middle East and North Africa (EMENA) and Latin America, but less well in the humid tropics of South Asia, and quite poorly in sub-Saharan Africa. However, the current competition for water in regions such as EMENA is becoming alarming. The historical performance may not continue without serious improvement of pre-planning, and the African projects need special care to ensure adequate future performance.

5.4 The Synoptic View of Water

The environmental interest groups outside the Bank and the Bank's own internal auditors felt seriously the conditions of water scarcity. The World Bank staff has realized the increasing costs of the water sector due to absence of a comprehensive view on it. Hence, it becomes imperative to look forward to a Bank-wide action taken policy in this sector.

Attempt has been made to take a synoptic view of the water problems in all its important uses and identify areas that require immediate attention/emphasis in the Bank's Policy Paper. This chapter has endeavoured to achieve a balance between discussing broad conceptual issues and specific sectoral issues. Water resources is a very broad subject; hence it is restricted to the discussion on the most important issues so that their inclusion in the Policy Paper can be justified. Mr Peter Rogers in his Working Paper (WPS 879) on "Comprehensive Water Resource Management – A Concept Paper" has opined that there are a few simple concepts and methodological approaches that could, if understood and adopted, significantly influence the successful development of water resources around the world. This attempt at simplification should be taken in the spirit of Occam's Razor exhaust the simplest explanation before requiring more complex ones.

5.5 Legal Issues, Water Markets, Privatization, Technology etc.

Discussions at length are made on (i) the legal issues of water rights, (ii) development of water markets, (iii) privatization of facilities, (iv) evolving of new technology, and (v) detailed sector-by-sector highlighting of problems and possibilities. To do justice to these issues, it is the contention of Mr Peter Rogers that the successful development of water resources lies in the refinement of economic analysis, the other issues are not a major focus. Hence there is a need to discuss some of the fundamental concepts underlying water management, and addresses the economic issues stemming from the fundamental concepts that cause problems in pursuing Bank water policy. The emphasis envisages on practical discussion of issues that are currently of widespread concern within the Bank and are likely to be of great importance into the next century.

5.6 Water Management – Fundamental Concepts

There are a few fundamental concepts that underlie the theory and practice of water management.

5.6.1 Water is a Unitary Resource

Although rain, surface water in rivers and lakes, groundwater, and polluted Freshwater are all part of the same resource base, but they have different manifestations, arising out of different parts of the hydrologic cycle. On the one hand, unutilized surface water is instrumental in recharging the groundwater; on the other hand, the over-pumped groundwater can narrow down the flows in the surface streams. Contaminated water can be purified by installing wastewater treatment processes or it can be recovered by the natural assimilative capacity of the surface and groundwater ecosystems. Thus, the hydrologic balance must be considered totally not just parts of it. Actions initiated in one part of the system often have significant impacts upon other parts of the system; these linkages must be considered while assessing the costs and benefits of specific actions.

5.6.2 Conflicts of Water Among Users

Even though water is a unitary resource, but there are often conflicts among users. The most common conflict lies between surface and groundwater users in irrigated agriculture. Improvement in the efficiency of application of the surface water results in an immediate decline in the availability of groundwater. Allocation decisions are also frequently ambiguous. For instance, water is often allocated to irrigation projects at low cost to the users when nearby urban areas are suffering serious water shortages and are willing to pay high prices for water. All these types of problems water use in all countries is proscribed by legal and institutional complexities. Many of these institutional complexities can create barrier in economically efficient use of water. These controls in the first place emerge because of the genuine problems that water raises for any politico-economic system. As the costs ignores the unitary nature of the resource rise, one would expect to feel the gradual easing of institutional constraints on development in the direction of more rational integrated use of the resource.

5.6.3 Scarcity of Water Associated with Its Concepts

The discussion about scarcity is usually intimately associated with the concepts of water as a renewable or non-renewable resource. If it is renewable, how can we ever run out of water? Certainly, in many arid or semi-arid zones, people are currently experiencing shortages of water. Part of this problem must do with the fact that under earlier historic conditions we were able to sustain the populations of cities and regions, even under relatively arid conditions. However, with the growth of population and income the demands for water seem to be overwhelming our technological capability to supply it. Under these conditions water scarcities and conflicts over water use begin to appear. The cost of supplying additional water to water-short areas is also increasing. Thus, it becomes increasingly difficult to supply the same amounts of water to users at the old low prices. The Malthus-Ricardo theory of resource scarcity seems to be working exactly as predicted: scarcity of resources limits economic growth, and ultimately brings it to a halt. It is no surprise that this theory earned economics the sobriquet of "the dismal science".

5.6.4 Economic Demand for Water

Modern economists define water scarcity slightly differently. According to them the need for water must be expressed in terms of a quantity and a price. This is called the "economic demand" for water and both quantity and price must be specified. In the current economic model's resources do not "run out". As the quantity demanded increases new sources will be tapped, at the same or higher costs, and/or the price will rise, restraining users' demand. This is essentially a self-limiting process. Clearly, if a resource base is fixed and the consuming population increases, something must give. This will be either the price or the quantity of the resource used. However, even this system would ultimately be forced to a halt by population or economic growth. The magical ingredient that enables the economic system to continue to grow is the existence, at some reasonable cost, of substitutes for the scarce resource. Although there is no substitute for water in sustaining human and animal life, there is an almost infinite supply of sea water, which can be converted at a cost of energy into Freshwater; then energy, or the capital to access the energy, becomes the limiting resource. Similarly, political boundaries, management skills, or human labour could limit the availability of water.

5.6.5 Treating Water Scarcity with Perceptions

Part of the problem of dealing rationally with water scarcity is the different perceptions that various groups have about how close we are to reaching the resource limits.

The World Bank in of their studies (Falkenburg et al. 1990) claimed that some countries (Israel, Jordan, Saudi Arabia, Syria, and Yemen) were unable to supply their populations with the minimum needed amount of water (500 cubic metres per capita per year). However, others seemed any possible adjustment mechanisms, which include recycling, reduction in agricultural use, changes in population policy, and the reclaiming of additional brackish water. Has water been a significant limit on economic growth in any of these countries? How will each country address the future of this resource? Has water been priced out of the reach of significant portions of the populations? What accounts for the fact that in many other countries with much larger per capita quantities of available water the per capita water use is lower? On closer inspection, the "water barrier" of 500 cubic metres per capita per year does not appear to be a real barrier in the Malthus-Ricardo sense. The idea that we are running out of water (Postel, 1990 and others) cannot be true globally, and even in specific water-short countries there is little doubt that water will always be available at some reasonable price, provided those countries follow sensible water policies. The definition of scarcity in unique economic terms is a distraction that can lead to major misallocations of the water resource.

5.7 Conclusion

The most useful economic literature on water is built on neo-classical economics. Much of it explains and deal best with aspects of "market failure", that is, when the classical model does not strictly apply. Eckstein (1958) mentions the following sources of market failure inherent in water resources as the most important:

Increasing returns-to-scale on the productions side are prevalent in water projects. For example, inland waterways and municipal water and wastewater services are natural monopolies because of the large economies of scale in the provision of the infrastructure. Many water-related investments tend to be very large to take advantage of these economies.

Externalities due to physical interdependence among production processes are inherent in many water activities. The externalities of both water quantity and water quality are experienced in the spatial sense between upstream and downstream users, and in a temporal sense between different seasonal releases of stored water, common pool effects on groundwater, and the export of pollution.

The classical model assumes that the income distribution in each setting is optimal. However, in development work it is rarely accepted that the income distribution in a country is the best one, and many water projects are specifically aimed at changing a mal-distribution of income.

When not all producers are small relative to the market, the marginality conditions for the existence of economically efficient solutions are violated. When government is involved it is often as the only producer in the market. In this case, the water supplied will make large changes in the local price of water, thus undermining the assumption of marginality inherent in benefit measurements.

The resources are not necessarily mobile. Typically, capital resources are relatively mobile but labour resources are not. Pockets of poverty and unemployment exist and many water projects (like the Tennessee Valley Project, TVA) were originally designed to address this lack of resource mobility. In addition, restricted water rights often impede the ease of transfer of water from one use to another.

Section 2

Integrated Water Resource Management

CHAPTER 6

Water: A Key Driver of Economic and Social Development

6.0 Introduction

Water is a key driver of economic and social development while it also has a basic function in maintaining the integrity of the natural environment. However, water is only one of many of vital natural resources and it is imperative that water issues are not considered in isolation.

6.1 Water Availability and Variability of Supply

There are great differences in water availability from region to region from the extremes of deserts to tropical forests. In addition, there is variability of supply through time as a result both of seasonal variation and inter annual variation.

All too often the magnitude of variability and the timing and duration of periods of high and low supply are not predictable; this equates to unreliability of the resource which poses great challenges to water managers and to societies. Most developed countries have, in large measure, artificially overcome natural variability by supply-side infrastructure to assure reliable supply and reduce risks, albeit at high cost and often with negative impacts on the environment and sometimes on human health and livelihoods. Many less developed countries, and some developed countries, are now finding that supply-side solutions alone are not adequate to address the ever-increasing demands from demographic, economic and climatic pressures; waste-water treatment, water recycling and demand management measures are being introduced to counter the challenges of inadequate supply. In addition to problems of water quantity there are also problems of water quality. Pollution of water sources is posing major problems for water users as well as for maintaining natural ecosystems.

6.2 Availability of Water in both Quantity and Quality

In many regions, the availability of water in both quantity and quality is being severely affected by climate variability and climate change, with precipitation in different regions and more extreme weather events. In many regions, too, demand is increasing as a result of population growth and other demographic changes (urbanization) and agricultural and industrial expansion following changes in consumption and production patterns. As a result some regions are now in a perpetual state of demand outstripping supply and in many more regions that is the case at critical times of the year or in years of low water availability.

6.3 Many Uses for Water

Water for basic human needs and reducing absolute poverty is directly related to the availability and quality of food and to the prevalence of disease. Clearly water is of fundamental importance for food production, for drinking, for sanitation and for hygiene. Adequate water in both quantities and quality underpins health and basic quality of life.

Water for social and economic development is clearly linked to the IWRM focus on the three E's- namely: equity, economics and environment. Water for social development includes the provision of education and health care.

6.4 Clean Water Supplies and Sanitation Helps Social Development

Without clean water supplies and good sanitation facilities in schools and hospitals social development is stymied. And for education – in schools without sanitation facilities – it is girls who suffer most and are therefore disadvantaged, introducing an important gender element into the equation. Water is of fundamental importance for economic development through energy and industrial production. It is needed for many forms of energy production – hydro power and the water for cooling of thermal and nuclear power stations. And energy in turn is needed for pumping, including extraction of water from underground aquifers. Water is needed for many industries and those industries in turn have effect, through pollution and abstraction, on water quality that affects both downstream users and natural ecosystems. A major water use is non-food agriculture, recent shifts towards growing bio-fuels. This has significant implications for water resources management.

6.4.1 Water and Natural Ecosystems

Natural ecosystems are of fundamental importance to human wellbeing and development. Our concern must not remain focussed on human development considerations only but it must place the human being, as an individual, as a member of a community and as part of society in an environmental context, to achieve well-being and harmony with nature. The loss of biodiversity and the degradation of ecosystems mean a loss of eco-system products and services and undermine the habitat Planet. Earth provides for humans. We destroy or degrade these natural systems at our peril, and so social and economic development and basic human betterment must go hand in hand with preservation of the natural environment.

6.4.2 Water Security

Floods, droughts, pollution spills into our water systems are of growing importance. Not only, in many regions, is there an increase in the frequency and intensity of floods, droughts and, with increasing industrialization, pollution spills, but, with increases in population, more people are living in zones prone to disasters. Also, with increased demand for scarce resources there is an increased risk of conflict over water: it is already part of the equation in many conflicts such as Darfur and the Middle East. Water security is also intrinsically linked to food security.

6.5 Diversity

While the world comprises many very different climatic and hydrological regions, which will be adversely impacted by climate change, there are many other aspects of diversity which affect the ways in which water is managed.

6.5.1 The Importance of Basin Management within the Context of Diversity

There is agreement among many that water should be managed within natural hydrological units – the river basin, lake basin or aquifer. However, geographic situations are diverse and natural units seldom coincide with administrative units. Some countries, such as Sri Lanka, are single national units in the sense that there are no international lands-borders with other countries. Indonesia is composed of any separate islands each of which has many river systems; administrative units may span both several islands and a large number of river basins. These examples contrast with such international river basins such as the Nile with the challenges associated with sharing the waters between upstream and downstream neighbours. A similar situation can also be seen within many large countries where rivers run through many states (Australia, China, India and USA). In other circumstances, such as those of the Rio Grande separating Mexico from the USA, the major river itself forms the boundary between nation states posing challenges for management of the resource. Some major aquifers also span national boundaries but as they are hidden their management is often neglected.

6.5.2 Diversity in Demographics

There are major contrasts in demographics between developed and developing countries. Many developing countries have very youthful populations virtually guaranteeing rapid population growth in the future; many developed countries by contrast have aging and diminishing populations. Simple growth or depletion in numbers is complicated by population movements. Urban populations are, in general, growing while rural populations are likely to grow at a much smaller pace or in some places diminish. There are also major migrations of population across international borders, some permanent, some seasonal and some, in the case of tourists, very short term; such population shifts intensify water management problems.

6.5.3 Diversity in Governance

Societies are organized in different ways from politically centralized to highly disperse; in some societies, such as federal jurisdictions, responsibilities for management of natural resources, including water, are primarily at provincial rather than at national level. Indeed, the availability of water was a major driver of the way governance structures developed. Currently, responsibilities for aspects of water management often are devolved to the community level even though they may have inadequate resources to undertake their responsibilities – this is often the case for drinking water supply, sanitation and hygiene.

6.5.4 Water Resources Reflect Cultural and Religious Beliefs

Attitudes of societies towards stewardship of water resources reflect cultural and religious beliefs and they differ greatly from country to country and often also within countries where populations are of diverse ethnic and social backgrounds. These differences are also manifested in the effectiveness and efficiency of institutions and of legislation. Financial resources and instruments so necessary especially in critical circumstances are often lacking in poorer societies.

It is not only governments, whether national, provincial or at lower levels of the municipality or community, that have responsibility in water management. Very often the private sector plays vital roles in the provision of water services. In many countries, public-private partnerships are being created to better manage supplies. Individual citizens, too have important roles to play, especially at the community level but all too often citizens do not have the means to express their demands and concerns. All these aspects of governance are critically important and affect the ability of societies to address their water challenges.

6.6 Fragmented to Integrated Management

As a rule, in the past with smaller populations, less intense economic activity and with less affluent societies demanding much less water, supply of the resource was usually much rear than demand for it. In such circumstances water for agriculture, for industry, for domestic and all other uses could be managed separately there being sufficient water to accommodate all needs and there being little competition between uses and between users. Moreover, water use by humans did not unduly impinge on the natural environment and ecosystems as it does today. Thus, it was common (and still is common) that within governments at both national and sub-national levels separate ministries would be set in place for each use for which water was needed.

6.6.1 Climate Change Adds Pressure on Limited Water Resources

As populations have grown, as food production has increased, as economic activity has developed and as societies have become more affluent, so demand for water has burgeoned. Climate change adds yet more pressure on our limited water resources. In very many places demand has far outstripped supply – this may be particularly so in seasons when supply may be severely limited or in years of drought, or at times when demand is particularly high, for example when there is great demand for water for irrigation.

6.6.2 Difficult Decisions on Water Allocations by Agencies

Thus managers, whether in the government, the private sector or local communities must make difficult decisions on water allocation? They find themselves in countries and regions that have very different physical characteristics and are at very different stages in economic and social development: hence here is a need for approaches to be tailored to the individual circumstance of country and local region.

More and more often managers must apportion diminishing supplies between ever-increasing demands considering the weaker voices of the poor and of the natural environment. The traditional fragmented or purely sectoral approach is no longer viable and a more holistic approach is essential.

6.7 Conclusion

This is the rationale for the Integrated Water Resources Management (IWRM) approach that has now been accepted internationally as the way forward for efficient and sustainable development and management of the world's limited water resources and for coping with conflicting demands. The most widely accepted definition of IWRM is that given by the Global Water Partnership: "IWRM is defined as a process that promotes the coordinated development and management of water, land and related resources, to maximize the resultant economic and social welfare in an equitable manner without compromising the sustainability of vital ecosystems".

There is a recognized need to develop a set of indicators which would characterize the status of implementation of the IWRM approach within countries. There have been many attempts to produce indicators which would adequately encompass diverse situations and the very different time scales at which implementation is taking place. The process is highly complicated and challenging. Moreover, this must be considered in the light of established reporting mechanisms, e.g. Statistics, and avoid adding onerous reporting demands on national governments.

UN-Water, has undertaken a major initiative through the World Water Assessment Programme to develop a comprehensive set of indicators – summary of progress is documented in the Second World Water Development Report.

To further develop suitable indicators UN-Water has established a Task Force on Indicators, Monitoring and Reporting. Many indicators already exist to measure social progress and the aim is to add value to these and not reinvent the wheel. A summary of progress made to date by the many agencies and organizations involved has been produced by UNEP-UCC. The Road-mapping initiative, being developed concurrently with this Report and complementary to it, lays out a timetable over the next seven years for the development of an achievable set of indicators including those specifically related to IWRM.

Reference

http://www.unwater.org.

CHAPTER 7

Objective Assessment of Global Freshwater Resources

7.0 The Response of the United Nations System

The need to set targets and monitor progress towards achieving the efficient and sustainable development and management of the world's limited water resources and for coping with conflicting demands. There is a well-recognized need to undertake comprehensive and objective assessments of the state of global freshwater resources. The water managers must address the uses to which the resources are put, the challenges associated with the resource and the ability of nations and societies to cope with the challenges. To this end, in the year 2000, the United Nations system created the World Water Assessment Programme (WWAP) with UNESCO leading the Programme by hosting its Secretariat. The WWAP has produced two World Water Development Reports (WWDRs) in 2003 and 2006. This process will continue to produce WWDRs every three years and thus provide a reporting mechanism to record the changes taking place in the resource itself and changing management challenges.

7.1 Need to Set Targets to Water Related Issues

It is also well recognized that there is a need to set targets towards which the world must strive if the many water-related challenges are to be resolved. Thus, in 2000, heads of State adopted the Millennium Declaration on the basis of which the UN instituted the Millennium Development Goals (MDGs). It can be argued that, to a greater or lesser degree, all the MDGs are water-related; with goal.

One related to growth and the others related to health or social issues. As a follow-up to the MDGs it was further agreed at the World Summit on Sustainable Development (WSSD) in Johannesburg in 2002, through the Johannesburg Plan of Implementation (JPoI), to "Develop integrated water resources management and water efficiency plans by 2005, with support to developing countries, through actions at all levels".

There was further discussion on IWRM and water efficiency plans at the CSD 12 and CSD 132 Meetings with a decision that at CSD 16 in 2008 there should be an assessment of progress made towards meeting the target. Comprehensive and systematic monitoring of all aspects of water resources and their management in an integrated fashion is undertaken by UN-Water through the WWAP; the series of WWDRs provide a reporting mechanism for the UN system.

7.2 Johannesburg Plan of Implementation (JPoI) and Integrated Water Resources Management (IWRM) Road Mapping Initiative

In association with the Johannesburg Plan of Implementation (JPoI) and Integrated Water Resources Management (IWRM) road mapping initiative has been started, facilitated by the Government of Denmark in collaboration with UN-Water, the Global Water Partnership and Representatives of Governments. This initiative recognizes the need for countries to set out "Roadmaps" that lay out a series of actions to be undertaken to apply an integrated approach to water resources development and management and to help meet the MDGs. It recognizes that different countries will need a set of actions suited to their needs and that time schedules for implementation would differ from country to country depending on specific country circumstances. In other words, solutions must be "tailor-made" or that "no one size fits all". The road-mapping initiative is being developed as a separate but complementary initiative to the current report.

7.3 Creation of the UN-Water Task Force on IWRM Monitoring and Reporting

In 2006, a Task Force on IWRM Monitoring and Reporting (TF) was created by UN-Water, with members drawn from UN-Water agencies and from partner organizations, with the mandate, inter alia, of producing the status Report on IWRM and Water Efficiency Plans for CSD6 (The Report).

The Report has been undertaken by UN-Water. The analyses within the Report draw primarily on the questionnaires undertaken by UN-DESA and UNEP (through the UNEP Collaborating Centre), during 2007 and supported by inputs from other members and partners of UN-Water, including UNDP, UN Statistics, WWAP and GWP. The Report also includes information gathered by the more informal surveys conducted by the Global Water Partnership (GWP) and the African Development Bank (AfDB).

For the purpose Report countries have been divided into two groups:
- Group 1 "developing" and "countries with economies in transition" (as defined by UN Statistics) and
- Group 2 "developed" (those belonging to either OECD or the European Union).

Regions and sub-regions are as defined by UN Statistics. Within the analyses more emphasis is placed on the countries with the greatest needs, i.e. those in Group 1.

Note 1: In addition to the IWRM target, a set of policy actions was adopted during the CSD meeting and UNDESA recently embarked on a study to assess the implementation of these actions for details see:
- http: //www. un. org/esa/sustdev/csd/csd
- 3_decision_unedited. pdf

Table 7.1:   Survey of Progress of IWRM

Countries responding to the UN-Water Survey (104 in total) and the surveys undertaken by GWP and the AfDB.

	Country * Least Developed Countires (2) Countries in transition	UN-Water Survey Response Y relates to the DESA questionnaire X relates to the UNEP questionnaire	GWP 2006 Survey 1=plan in place 2=plans in preparation 3=only initial steps taken	AfDB Survey
DEVELOPING COUNTRIES				
AFRICA				
East Africa	Burundi*		3	
	Djibouti*		3	
	Entrea*	Y	2	
	Ethiopia*		2	
	Kenya		2	X
	Malawi*	Y	2	
	Mauritius	X	2	
	Mozambique*	X	2	
	Rwanda*		3	X
	Seycheles	Y		
	Tanzania*	X	2	
	Uganda*	Y	1	
	Zambia*	X	2	
	Zimbabwe	X	1	
Central Africa	Angola*	X	3	
	Cameroon		2	X
	Central African Rep*		3	X
	Chad*		3	

	Congo		3	
	DR Congo*	X	3	X
Northern Africa	Algeria	X	3	
	Egypt	Y	2	X
	Libya	X	3	
	Morocco	X	2	
	Sudan*		2	
	Tunisia	Y	2	X
Southern Africa	Botswana	X	2	
	Lesotho*	Y	3	
	Namibia	Y	1	
	South Africa	X	1	
	Swaziland	X	2	
Western Africa	Benin*		2	X
	Burkina Faso*	Y	1	X
	Cape Verde*	Y	3	X
	Cote d'Ivoire	X		X
	Ghana	Y	2	
	Guinea*	Y		X
	Liberia*	Y		X
	Mali*		2	
	Mauritania*	X	2	X
	Niger*			X
	Nigeria		2	
	Senegal*		2	X
	Seirra Leone*	Y		
	Togo*	Y		X
AMERICAS				
Caribbean	Anguilla	X		
	Antigua and Barbuda	X		
	Bahamas	X		
	Barbados	Y	2	

	Cuba	Y		
	Dominica	X		
	Grenada	X		
	Jamaica	Y	2	
	Montserrat	X		
	Saint Kitts and Nevis	Y		
	Saint Lucia	X		
	Trinidad and Tobago		2	
Central America	Belize	X	2	
	Costa Rica	Y	2	
	El Salvador	X	2	
	Guatemala	Y	3	
	Honduras	X	3	
	Nicaragua	X	2	
	Panama	X	2	
South America	Argentina	Y	2	
	Bolivia	X	3	
	Brazil	X	1	
	Chile	X	2	
	Colombia	Y	2	
	Ecuador	X		
	Paraguay	X	3	
	Peru	X	2	
	Uruguay	X	2	
	Venezuela	X	3	
ASIA				
Central Asia	Kazakhstan(2)	Y	1	
	Kyrgyzstan(2)	Y	2	
	Tajikistan(2)	Y	2	
	Turkmenistan(2)	Y	2	
	Uzbekistan(2)	Y	2	
Eastern Asia	China	Y	1	

Southern Asia	Bangladesh*		1	
	India		2	
	Nepal*		2	
	Pakistan		2	
	Sri Lanka	Y***	3	
South Eastern Asia	Cambodia*	Y	3	
	Indonesia	X	2	
	Lao Peoples's DR*	X	2	
	Malaysia		2	
	Myanmar*		3	
	Philippines	Y	2	
	Thailand	X	1	
	Vietnam	Y	3	
Western Asia	Armenia (2)	Y	1	
	Azerbaijan (2)	Y	3	
	Georgia (2)	Y	3	
	Jordan	Y		
	Syrian Arab Republic	Y		
OCEANIA				
Melanesia	Fiji		2	
	Solomon Islands*		3	
Micronesia				
Polynesia	Samoa*		1	
	Tuvalu*		3	
EUROPE				
Southern Europe	Croatia (2)	Y		
	Serbia (2)	Y		
DEVELOPED COUNTRIES				
Asia	Japan	Y		
	Republic of Korea	Y		
	Turkey	Y		

Northern America	USA	Y		
Central America	Mexico	Y		
Eastern Europe	Cyprus	Y		
	Bulgaria		2	
	Czech Republic	Y	1	
	Hungary	Y	1	
	Poland		1	
	Romania	Y	1	
	Slovakia		1	
Northern Europe	Denmark	Y		
	Estonia	Y	1	
	Finland	Y		
	Ireland	Y		
	Latvia	Y	1	
	Lithuania		2	
	Norway	Y		
	Sweden	Y		
	Greece	Y		
	Malta	Y		
	Portugal	Y		
	Slovenia		2	
	Spain	Y		
Western Europe	Austria	Y		
	France	Y		
	Germany	Y		
	Netherlands	Y		
	Switzerland	Y		
Oceania	Australia	Y	1	
	New Zealand	Y		

***Sri Lanka is not included in the analysis as it did not respond to the official UN-DESA questionnaire even though it did respond to a trial run for the questionnaire.

Table 7.2:   Summary Statistics for Country Survey

	Region and Sub-region	UN-Water Survey 2007	GWP 2006 Survey	AfDB Survey
AFRICA	Eastern Africa	9	13	2
	Middle Africa	2	6	3
	Northern Africa	5	6	2
	Southern Africa	5	5	0
	Western Africa	9	8	10
	Total	30	38	17
AMERICAS	Caribbean	11	3	
	Central America	7	7	
	Southern America	10	9	
	Total	28	19	
ASIA	Central Asia	5	5	
	Eastern Asia	1	1	
	Southern Asia	0	5	
	South-Eastern Asia	6	8	
	Western Asia	5	3	
	Total	17	22	
EUROPE	Eastern Europe	0	0	
	Southern Europe	2	0	
	Total	2	0	
OCEANIA		0	5	
Total developing countries		77	84	
	developed countries	27	11	
Grand total		104	95	

7.4 Comments on the Surveys

The questionnaires were addressed to governments at the national level. Therefore, they do not reflect responsibilities for management at sub-national levels. The case studies demonstrate that many management decisions are made at the provincial and community levels.

The GWP and AfDB surveys were more informal and are useful as they reflect the views of a different set of stakeholders and therefore provide an alternative perspective.

The UN-DESA questionnaire:

- 27 developed countries and 39 developing countries (including countries with economies in transition) responded. Of the 39 developing countries that responded, 7 responded through UNEP (Burkina Faso, Cape Verde, Guinea, Liberia, Sierra Leone, Togo and Uganda).

- A total of 65 questions were posed to be answered in multiple choice fashion; a further 8 questions allowed written answers to elaborate in more detail.
- The responses to the 65 questions are found in the Database, Worksheet 2: Responses to UN-DESA questionnaire; the responses to the 8 written answers may be accessed directly through the same Worksheet for specific countries.
- There are many cases where countries, in answering the questionnaire, have ticked more than one box on the same line. In such a case, UN-DESA, in making the initial compilation of the responses, has elected to select just one answer as the most reasonable choice.
- Many countries have chosen not to answer all the questions. The summary statistics simply ignore these omissions.

The UNEP questionnaire:
- A total of 58 countries responded to the UNEP questionnaire; the complete set of responses is found in the Database, Worksheet 3: Responses to UNEP questionnaire.
- For 17 countries, there are responses to both the UN-DESA and the UNEP questionnaires; this allows an inter-comparison of responses which is important in assessing their compatibility. The information for the inter-comparison is found in the Database, Worksheet 5: DESA-UNEP comparison.

Merging of the UN-DESA and UNEP questionnaires:
- The information for the 39 developing countries within the UN-DESA questionnaire has been supplemented for 38 additional countries by partial responses from similar questions in the UNEP questionnaire. Of the 65 questions posed by UN-DESA 26, had exact or very similar counterparts in the UNEP questionnaire. Overall the answers to the UNEP questionnaire are slightly lower than the answers to the UN-DESA (for 18 questions they are lower and for 7 questions they are higher).
- The summary statistics have been prepared from the responses from 77 developing countries plus responses from 27 developed countries.
- There are contrasts in the responses from different regions. Apart from a partial response from Sri Lanka, there are no responses at all from South Asia – a major gap in the survey. In contrast there is a complete set of responses from Central Asia.

The GWP Survey

This survey covered 95 countries, 84 developing and 11 developed. For 59 of these countries data from the UN-Water Survey are also available allowing a valuable inter-comparison between these in formal and official surveys.

The AfDB Survey

This survey covered 17 countries in Africa; the survey questions were a direct sub-set of the UNEP questionnaire. This survey is used to supplement the other 3 surveys within the African context.

CHAPTER 8

Integrated Water Resources Management Planning

8.1 Analysis of the UN Water Survey

Care must be taken in the analysis of the questionnaires sent out by UN-DESA and UNEP for the following reasons:

- It must be recognized that many of the very poorest countries were unable to respond to the questionnaires through lack of capacity to do so; conversely a larger proportion of developed countries than developing countries did respond to the survey. In this sense, the survey is biased towards countries more capable of giving responses.
- Some regions of the world, particularly South Asia, are under-represented as responses from many of the countries concerned were not forthcoming – in this sense there is regional bias.
- In surveys of this type there is always room for differing interpretation of the meaning of questions because of cultural and linguistic diversity; indeed this may result in more "optimistic" interpretation of situation and status by some countries than by others.
- This survey was aimed primarily at national governments. In many countries responsibility and authority for water management, especially in federal jurisdictions, is subordinated to sub-national levels; conversely some national governments must manage their water within a broader context of international river basins or of regional jurisdictions, for example in the case of the EU where the European Framework Directive becomes more important than purely national plans and policies.
- Several of the questions are not relevant to all countries; for example, trans-boundary water issues may not be relevant to Small Island countries, humid regions may not be concerned with questions of aridity and land-locked countries are unlikely to be concerned with desalination.
- Despite these caveats, it is still possible to discern overall trends and to draw several broad conclusions from the survey.
  Comparative results between major country groupings and between regions and sub-regions are presented in Tables 8.1a and 8.1b. These comparisons are provided in the diagrams (Figures 8.1, 8.2 and 8.3).

Table 8.1a: Comparison of Developed Countries with Africa, the Americas and Asia

Main national instruments and other national/federal strategies that may contribute to promoting IWRM	• Developed countries significantly more advanced on main national instruments. • Asia and the Americas more advanced on national development plans and national environmental action plans with IWRM components. • Of developing countries Africa, least advanced with poverty reduction strategies with WRM components.
Water resources development	• Developed countries more advanced on most issues, but, as expected, not for rain-water harvesting. • Asia more advanced than other developing regions for WR assessment.
Water resources management	• Developed countries significantly more advanced except in the less relevant areas of combating desertification and irrigated agriculture. • Developing regions very similar except the Americas more advanced in programmes and policies for watershed management, groundwater. management and drainage and irrigation; Asia more advanced in legislative mechanisms to control pollution.
Water use	• Developed countries significantly more advanced. • Africa consistently less advanced than other regions.
Monitoring, information management and dissemination	• Developed countries significantly more advanced. • Asia more advanced than the Americas which in turn are more advanced than Africa on all issues except monitoring and reporting the impacts of IWRM reforms where Africa is more advanced.
Capacity building and enabling environment	• Developed regions significantly more advanced on all issues except pro-poor policies which are designated not relevant by many developed countries. • Similar responses from developing regions with some interesting contrasts – e.g. Asia more advanced on institutional reforms yet behind on institutional coordination mechanisms.

Source: Table 3a: Sub-regional comparisons, on the "Status Report on Integrated Water Resources Management and Water Efficiency Plans" prepared for the 16th session of the Commission on Sustainable Development – May 2008.

Table 8.1b: Sub-Regional Comparisons

Particulars	Africa	America	Asia
Main national instruments and other federal strategies that may contribute to promoting IWRM	Countries of N Africa score significantly higher on main national instruments while countries of S Africa score higher on plans with IWRM components and on sustainable development strategies.	Major differences between Caribbean countries and countries of S America – Caribbean much higher on main national instruments; S America much higher on other plans contributing to IWRM.	W Asia: generally low scores all round. SE Asia high on national/federal IWRM and water efficiency plans in contrast to Central Asia.

Particulars	Africa	America	Asia
Water resources development	Note: several issues (e.g. desalination and coastal fog harvesting) not relevant to many countries – otherwise very similar responses.	Similar responses except for Caribbean countries which rank high for assessment, regulatory norms and basin studies but low on recycling.	Here a definite consistent hierarchy of responses from China with highest scores through SE Asia, W Asia to Central Asia.
Water resources management	Very similar responses overall except for N Africa which, as would be expected in arid environments, has higher scores on groundwater, certification and irrigation issues.	A consistent hierarchy of scores Caribbean being consistently highest (except as expected, in shared management of resources); Central Americas being consistently lowest.	A consistent hierarchy of responses with E and SE Asia having high scores and Central Asia having low scores.
Water use	Northern Africa consistently higher scores than other African regions which display similar responses.	Caribbean countries have significantly higher scores than other regions of the Americas.	E and SE Asia consistently higher than Central and W Asia.
Monitoring, information management and dissemination	N Africa consistently higher scores than other African regions which display similar responses.	Caribbean highest on most measures.	Central Asia generally has lowest scores.
Capacity building and enabling environment	Similar responses – N Africa highest on some responses.	Caribbean generally highest.	Caribbean generally highest.
Stakeholder participation	N Africa generally highest, E Africa lowest.	Central America generally low scores.	E and SE Asia generally high; Central Asia lower.
Financing	N Africa generally higher scores; not many differences for other African regions.	S America generally with highest scores except for Caribbean with gradual cost recovery mechanisms and strategies.	SE Asia generally with the highest scores.

Source: Table 3b: Sub-regional comparisons, on the "Status Report on Integrated Water Resources Management and Water Efficiency Plans" prepared for the 16th session of the Commission on Sustainable Development – May 2008.

8.2 Comparative Analysis of the UN Water Survey with Global Water Partnership, (GWP) and African Development Bank (AfDB)

The purpose of this comparison of surveys was an attempt to assess progress towards putting IWRM plans in place. The GWP Survey was carried out about 18 months before the UN-Water Survey, therefore only small changes might be expected as this is a relatively short-time period.

The GWP Survey was carried out at the end of 2005 specifically to assess the extent to which the WSSD target has been met. Thus, it focussed on the creation of IWRM plans and did not assess the extent of implementation of plans. The GWP Survey evaluated 95 countries (11 of which were developed countries having high scores) and concluded that:

- 20 countries (21 per cent) had plans/strategies in place or a process well underway, and that incorporated the main elements of an IWRM approach.
- 50 countries (53 per cent) were in the process of preparing national strategies or plans but require further work to live up to the requirements of an IWRM approach.
- 25 countries (26 per cent) had taken only initial steps in the process towards preparing national strategies or plans and had not yet fully embraced the requirements of an IWRM approach.

Fifty-nine countries (Africa-24; Americas-14; Asia-15; Developed countries-6) are covered by both the GWP and UN–Water Surveys. Although the questionnaires used for the GWP Survey and UN-Water Survey are not completely comparable and use different terminology they are sufficiently like enable general comparisons to be made.

To make a comparative analysis of results for the informal GWP Survey and the official UN-Water Survey, the order of the original GWP Survey classification has been reversed so that responses are ranked in ascending order from least to most advance.

Table 8.2: The Ranking for the GWP and UN-Water Surveys

UN Water Survey	GWP Survey (order reversed)	Comments
Not relevant	Countries that have taken only initial steps in the process towards preparing national strategies/plans and have not yet fully embraced the requirements of an IWRM approach.	—
Under consideration	Countries that are in the process of preparing national strategies plans but require further work to live up to the requirements of an IWRM approach.	—
In place but not yet implemented	Countries that have plans/strategies in place, or a process well under way, and that incorporate the main elements of an IWRM approach.	For comparison purposes, all those countries included in categories 3, 4 and 5 of the UN Water survey also satisfy category 3 of the GWP survey.
In place and partially implemented	Not assessed	
Fully implemented	Not assessed	

Source: Table 4: Sub-regional comparisons, on the "Status Report on Integrated Water Resources Management and Water Efficiency Plans" prepared for the 16th session of the Commission on Sustainable Development – May 2008.

Table 8.3: Summary Statistics for GWP and UN-Water Surveys

Region	Number of countries	GWP category 3		GWP category 2		GWP category 1		UN-Water category 3		UN-Water category 2		UN-Water category 1	
		No	%	No	%	No	%	No	%	No	%	No	%
E Africa	8	2		6		0		3		5		0	
Central Africa	2	0		0		0		2		0		0	
N Africa	5	0		3		2		2		3		0	
S Africa	5	2		2		1		2		3		0	
W Africa	4	2		2		0		2		2		0	
Africa total	**24**	**6**		**13** **5.5**	**54.2**	**5**	**20.8**	**9**	**37.5**	**13**	**62.5**	**0**	**0.0**
Caribbean	2	0		2		0		2		0		0	
Central Americas	5	0		3		2		1		4		0	
S America	7	1		4		2		3		4		0	
Americas total	**14**	**1**	**7.1**	**9**	**64.3**	**4**	**28.6**	**6**	**42.9**	**8**	**57.0**	**0**	**0.0**
Central Asia	5	1		4		0		0		1		4	
E Asia	1	1		0		0		1		0		0	
SE Asia	6	1		3		2		4		2		0	

Region	Number of countries	GWP category 3		GWP category 2		GWP category 1		UN-Water category 3		UN-Water category 2		UN-Water category 1	
		No	%	No	%	No	%	No	%	No	%	No	%
W Asia 2	3	1		0		2		0		1		2	
Asia total	15	4	26.7	7	46.7	4	26.7	5	33.3	4	26.7	6	40.0
Developing countries total	53	11	20.8	29	54.7	13	24.5	20	37.7	27	50.9	6	11.3
Developed countries	6	6	100	0		0			100	0		0	

Source: Table 5: Sub-regional comparisons, on the "Status Report on Integrated Water Resources Management and Water Efficiency Plans" prepared for the 16th session of the Commission on Sustainable Development – May 2008.

The major conclusions from these listings are as follows:
- Developed countries
  For the six countries considered in this comparison there are no significant differences between the surveys; as a group the developed countries are well advanced in the process of incorporating IWRM principles into their national plans and most are well on their way to implement those plans.
- Developing countries and countries with economies in transition
  1. For the 53 countries considered in this comparison there are modest but significant improvements in the summary statistics:
  2. In 22 countries the UN-Water Survey shows a higher level of progress than the GWP Survey;
  3. While in 7 countries there seems to have been a lower level of progress (6 of these being in Asia);
  4. In 24 countries there has been little measurable change;
  5. It is in the Americas that the greatest overall progress has been made.

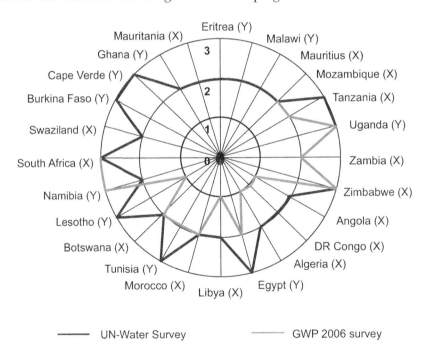

── UN-Water Survey   ┄┄ GWP 2006 survey

Figure 8.1:   GWP UN–Water Comparison for Africa

- Progress from only initial steps to plans in preparation or in place in Angola, DR Congo, Algeria and Libya.
- Progress from only initial steps to plans in place in Lesotho.
- Progress from plans in preparation to plans completed and/or under implementation in Tanzania, Egypt and Tunisia.
- Decline from plans in place to only in perparation in Namibia.

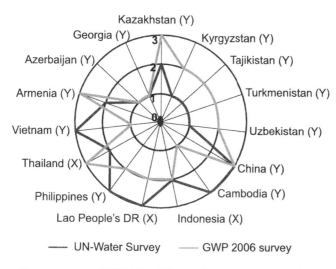

Figure 8.2:   GWP UN–Water Comparison for Asia

Source: "Status Report on Integrated Water Resources Management and Water Efficiency Plans" prepared for the 16th session of the Commission on Sustainable Development – May 2008.

- Progress from only initial steps to plans completed and/or under implementation in Cambodia and Vietnam.
- Progress from plans in preparation to plans completed and/or under implementation in Lao People's Republic and Philippines.
- Decline from plans in place to only in preparation in Thailand, Kazakhstan and Armenia; from in preparation to no steps taken in Kyrgyzstan, Tajikistan, Turkmenistan and Uzbekistan.

Figure 8.3:   GWP UN–Water Comparison for Developed Countries Czech Republic (Y)

All developed countries static or making progress.

8.3 Modest Progress

Given the relatively short time between the surveys it would be expected that only some modest progress would be made and this is confirmed by these figures. In individual cases the GWP or UN-Water Survey may be optimistic. For example, GWP results seem overly optimistic for Central Asia. The UN-Water results may also be more optimistic as they are completed by officials who may be inclined to give a more positive result. Also, the questionnaires may not be sufficiently robust to capture nuances in understanding by different cultures and language groups and interpretation of the questions may well be subject to individual bias of those filing the answers.

8.4 Conclusion

Nevertheless, overall the results are sufficiently similar overall to conclude that the two surveys are comparable and indicate some progress since 2005. The AfDB undertook an additional survey in 2007; it covered 17 African countries. Six of these countries were not covered by the UN-Water Survey but were included in the GWP Survey: Benin, Cameroon, Central African Republic, Kenya, Rwanda and Senegal. All these countries fall into the GWP categories of either being in the very initial stages of developing national plans or the plans have yet to properly incorporate IWRM principles. The UN-Water Survey shows no significant progress made in these countries.

CHAPTER 9

Implementation of IWRM and Water Efficiency Plans

9.0 Introduction

The purpose of this section is an attempt to assess the extent to which countries have been able to go beyond simply having plans in place to the stage of implementing those plans and the extent to which tangible outcomes have been forthcoming.

9.1 Response of UN-Water Survey

Table 9.1 presents responses to the UN-Water Survey on the questions of the extent to which countries have implemented IWRM and Water Efficiency Plans. It focusses on those countries that have plans in place and which are either partially or fully implemented.

Table 9.1: Response to Questions 6, 8b and 8c

Particulars	Question 6, If the country is in the stage of implementation, indicate specific actions taken.		Question 8b, What are the main water management measures undertaken?		Question 8c, What are the results achieved?	
Developing countries (37)						
Several specific actions taken	11		Several measures taken	10	Good results achieved	7
Some actions taken	13		Some measures taken	21	Some results achieved	19
No actions taken	–		No measures taken	1	No results achieved	4
No response	3		No response	5	No response	7
Developed countries (27)						
Several specific actions taken	25		Several measures taken	20	Good results achieved	10

Some actions taken	1	Some measures taken	6	Some results achieved	13
No actions taken	–	No measures taken	1	No results achieved	2
No response	1	No response	–	No response	2

Source: Table 7 Sub-regional comparisons, on the "Status Report on Integrated Water Resources Management and Water Efficiency Plans" prepared for the 16th session of the Commission on Sustainable Development – May 2008.

9.2 Examples of Ongoing IWRM Processes

Table 9.1 provides examples of developing countries that have found IWRM a useful framework for management of water resources and have included it as a pivotal concept.

The concept has been included in key Government documents that guide and regulate the use, conservation and protection of a nation's water resources and implementation at local level is ongoing. The table is not exhaustive.

In addition to what is documented here, there are many ongoing and planned IWRM programmes; as well as numerous national and regional IWRM partnerships and initiatives related to transboundary waters.

Table 9.2: Evidence of Adoption and Use of the IWRM Approach

Eritrea	• Integrated Water Resources Management and Water Efficiency Plan (IWRM/WE) – Ministry of Land Water and Environment (draft 2007)
Malawi	• National Water Policy – Ministry of Irrigation and Water Development (2005) • Water Resources Act No. 15 of 1969 with later amendments. Government of Malawi • Integrated Water Resources Management/Water Efficiency (IWRM/WE) Plan for Malawi – Ministry of Irrigation and Water Development (draft 2007)
Mozambique	• National Water Resource Strategy – Department of Water Affairs and Forestry (2004) • IWRM Plan – National Directorate of Water Affairs (draft 2007)
Seychelles	• Water Regulations – Public Utilities Corporation (1988) • Water Supply Development Plan – Public Utilities Corporation (2005) • Water Policy – Public Utilities Corporation
Tanzania	• National Water Sector Development Programme 2006–2025 – Ministry of Water (2006) • IWRM Strategy and Action Plan – Ministry of Water (2004) • National Water Policy – Ministry of Water (2002) • National Water Law based on revised Water Act No. 42 of 1974 – Government of Tanzania (draft 2007)

Uganda	• A National Water Policy – Ministry of Water, Lands and Environment (1999) • National Water Action Plan – Water Resources Management Department (1994) • Water Resources Management Reform Strategy – Water Resources Management Department (2005) • National Water Quality Management Strategy – Ministry of Water and Environment (2006)
Zambia	• IWRM and Water Efficiency Plan – Ministry of Energy and Water Development (2006) • The Revised National Water Policy – Ministry of Energy and Water Development (2007) • Water Resources Management Bill – Ministry of Energy and Water Development (draft 2007) • National Development Plan – Ministry of Energy and Water Development (2007)
Angola	• IWRM and Water Efficiency Roadmap – Ministry of Water and Energy (draft 2007)
Algeria	• National Plan for Water – Ministry of Water Resources (2003) • National Water Law – Government of Algeria (2005) • Action Plan for the implementation of an IWRM Framework – Ministry of Water Resources (draft 2006–2007)
Egypt	• National Water Resources Plan – Ministry of Water Resources and Irrigation (2004)
Morocco	• Master Plans of Integrated Water Resources Development for River Basins – Ministry of Land, Water and Environment (2001) • National Water Plan – Ministry of Land, Water and Environment (2006) • Decree no 2–05–1594 – Development and Revision of Master Plans and National Plans for Integrated Water Resources Management – Government of Morocco
Tunisia	• The Water Code (Law no. 16) – Ministry of Agriculture and Water Resources (1975) • Water Master Plan for the North of Tunisia – Ministry of Agriculture and Water Resources (1970) • Water Master Plan for the Centre of Tunisia – Ministry of Agriculture and Water Resources (1977) • Water Master Plan for the South of Tunisia – Ministry of Agriculture and Water Resources (1983) • Water Resources Mobilization Strategies – Ministry of Agriculture and Water Resources (1990) • Water Conservation Strategy – Ministry of Agriculture and Water Resources (1995)
Botswana	• IWRM Strategy and Action Plan – Ministry of Minerals, Energy and Water Resources (2006)
Lesotho	• Roadmap to completing integrated water resources management and water efficiency planning in Lesotho – Ministry of Natural Resources, Water Commission (April 2007)

Namibia	• National Water Policy White Paper – Government of Namibia (2000) • Water Resources Management Act – Government of Namibia (2004) • Integrated Water Resources Management Strategy and Action Plan – Ministry of Agriculture, Water and Rural Development (2006)
Swaziland	• Water Policy – Ministry of Natural Resources and Energy (draft 2007) • IWRM and Water Efficiency Plan – Water Resources Branch (draft 2007) • Water Act (2003) – Government of Swaziland
Burkina Faso	• Decree No. 2003–220: Action Plan for IWRM in Burkina Faso (PAGIRE) – Ministry of Agriculture, Hydraulics and Fishing Resources (2003) • Burkina Faso Water Vision – Ministry of Agriculture, Hydraulics and Fishing Resources (2000) • Water Law No.002-2001 – Government of Burkina Faso (2001)
Cote d'Ivoire	• IWRM Roadmap 2007-2015 – Ministry of Environment, Water and Forestry (2007)
Ghana	• IWRM Component Support programme (2004–008) – Water Resources Commission (2004) • Water Resources Policy – Water Resources Commission (draft 2007)
Liberia	• Liberia IWRM Roadmap – Ministry of Lands, Mines and Energy (draft 2007) • National Water Policy – Ministry of Lands, Mines and Energy (draft 2007)
Mauritania	• IWRM Action Plan – National Council for Water (2007) • National Development Policy for Water and Energy – Ministries of Water, Energy and Environment (1998) • National Water Act (Article 3) – Government of Mauritania (2005)
Togo	• National Water Policy – Directorate of Water and Sewerage (draft 2007) • National Water Law – Directorate of Water and Sewerage (draft 2007) • IWRM Roadmap – Directorate of Water and Sewerage (draft 2007)
Barbados	• National Water Resources Management and Development Policy – Government of Barbados (draft 2002) • National Water Law – Government of Barbados • Marine Pollution Control Act – Government of Barbados (1998) • Emergency Drought Management Plan – Government of Barbados (1998) • IWRM and Water Efficiency Plan – In place and partially implemented
Cuba	• National Water Policy – Ministry of Science, Technology and Environment (2000) • National Water Strategy – Ministry of Science, Technology and Environment (2000) • Water Conservation and Efficient Use Strategy – Ministry of Science, Technology and Environment (2005) • National Environment Management Strategy – Government of Cuba (2007)
Grenada	• Simultaneous preparation of IWRM Roadmap and National Water Policy – Water Policy Steering Committee (April 2007)

Jamaica	• Water Resources Act – Government of Jamaica (1995) • National Water Policy, Strategy and Action Plan – Government of Jamaica (1999) • National Water Resources Development Master Plan – Government of Jamaica (1990) • National IWRM Framework – Water Resources Authority (2001)
Costa Rica	• National Strategy for Integrated Water Resources Management – Government of Costa Rica (2006) • National IWRM Action Plan – Government of Costa Rica (2006) • National Water Law (No. 14585) – Government of Costa Rica (draft 2006)
Guatemala	• National Water Policy – Ministry of Environment and Natural Resources (2004) • National Water Law (Initiative 3118) – Ministry of Environment and Natural Resources (2005) • Plan for the Sustainable Use and Management of Water Resource (Initiative 3419) – Ministry of Environment and Natural Resources (2005) • National Law for the Protection of River Basins (Initiative 3317) – Ministry of Environment Natural Resources (2006) • National iwrm Policy – Government of Guatemala (2006) • National iwrm Strategy – Government of Guatemala (2006) • Environment and Natural Resource Conservation Policy – Government of Guatemala (2007)
Honduras	• IWRM Action Plan – Honduran Water Platform (2006)
Nicaragua	• General Law on National Waters – Government of Nicaragua (2007) • Environmental Action Plan – Ministry of Environment (1994) • IWRM Action Plan – Ministry of Environment (1996)
Argentina	• IWRM Roadmap – Sub-secretariat of Water Resources (2007)
Brazil	• National Water Policy (Law No. 9433) – Government of Brazil (1997) • National Water Resources Plan – Ministry of Environment (SRH/MMA), National Water Council (CNRH) and National Water Agency (ANA) (2007)
Colombia	• National Development Plan 2006–2010 – National Planning Department (2006)
Kazakhstan	• IWRM National Roadmap including proposed project outlines – speed-up of the IWRM 2005 objectives implementation in Central Asia – Government of Kazakhstan (2006)
China	• National Water Law – (2002) • Water Pollution Prevention and Control Law – (1996) • National Flood Control Law – (1997) • National Water and Soil Conservation Law – (1991) • IWRM Plan – Planning process initiated in 2002 and still ongoing
Cambodia	• Integrated Water Resources Management (IWRM 2005) and Roadmaps in Cambodia – Department of Water Resources Management and Conservation (2006) • Water Law – Royal Government of Cambodia (Sept. 2006)

Indonesia	• National Water Law No.7/2004 – Government of Indonesia (2004) • IWRM Roadmap – Directorate General Water Resources of Ministry of Public Works (2006)
Lao PDR	• Policy on Water and Water Resources – Government of Lao PDR (draft 2000) • The Law on Water and Water Resources – Government of Lao PDR (1996) • IWRM National Roadmap – Water Resources Co-ordination Committee Secretariat (2006)
Malaysia	• 9th Malaysia Plan – Economic Planning Unit – Prime Minister's Department (2006) • National Study for the Effective Implementation of IWRM in Malaysia – Ministry of Natural Resources and Environment (2006) • Our Vision for Water in the 21st Century – Ministry of Natural Resources and Environment (2000)
Philippines	• Medium Term Philippine Development Plan (2004-2010) – Government of Philippines (2004) • Clean Water Act – Government of Philippines (2004) • Integrated Water Resources Management (IWRM) Plan Framework – National Water Resources Board (2007)
Thailand	• National Water Law/Code – Government. of Thailand (draft 2007) • National Water Policy – Ministry of Natural Resources and Environment (2000) • IWRM National Roadmap – Department of Water Resources (2007)
Vietnam	• Law on Water Resources – Government of Vietnam (1998) • National Water Resources Strategy- Government of Vietnam (2006) • National Strategy on Rural Clean Water Supply and Sanitation – Government of Vietnam (2000) • National Strategic Program of Action On Desertification Control – Government of Vietnam (2006) • IWRM and Water Efficiency Plan – In place and partially implemented
Armenia	• Water Code – Government of Armenia (2002) • National Water Policy – Government of Armenia (2005) • National Water Programme – Government of Armenia (draft 2007)
Azerbaijan	• Law of Azerbaijan Republic on Amelioration and Irrigation – Azerbaijan Republic (1996) • Water Code Azerbaijan Republic – Azerbaijan Republic (1997) • Law of Azerbaijan Republic on Water Supply and Water Drainage System – Azerbaijan Republic (1999) • Law of Azerbaijan Republic on Municipality Water Resources Management – Azerbaijan Republic (2001) • National Program of Development of Amelioration and Water Resources Management of Azerbaijan (2007–2015) – Azerbaijan Republic (2006)
Jordan	• National Water Policy – Ministry of Water and Irrigation • National Water Strategy – Ministry of Water and Irrigation (2003) • National Water Master Plan – Ministry of Water and Irrigation (2004)

Syria	• National Water Policy – Government of Syria • National Water Law (No. 31) – Government of Syria (2005) • IWRM and Water Efficiency Plan – In place but partially implemented
Croatia	• Water Act (OG 107/95, 150/05) – Ministry of Agriculture, Forestry and Water Management Financing Act (OG 107/95, 19/96, 88/98, 150/05 – Ministry of Agriculture, Forestry and Water Management • National Environmental Strategy with embodied National Action Plan (NEAP) (OG 46/02) – Ministry of Agriculture, Forestry and Water Management • IWRM and Water Efficiency Plan – Under consideration
Serbia	• National Water Policy • National Water Law – Ministry of Agriculture, Forest and Water Management (1991) • Enviromental Protection Law – Ministry of Agriculture, Forest and Water Management (2004) • Water Resource Management Master Plan – Ministry of Agriculture Forestry and Water Management (2002)

Sources: http://www.gwptoolbox.org, http://www.unesco.org/water/wwap/wwdr/wwdr2

9.3 Case Studies from Selected Countries

It is instructive to cite specific examples of the implementation of the IWRM approach and the benefits there by to be derived. While the UN-Water Survey was aimed primarily at the national level, countries sharing river basins must also consider trans-boundary implications and include them in their planning; conversely many actions must be taken at sub-national and at very local levels to manage water wisely. The examples below cover a range of circumstance and are illustrative of the diversity of situation with a multiplicity of beneficial outcomes.

IWRM in action at the local level, as well as national and international level, is illustrated in detail in more than 200 case studies within the Global Water Partnership's IWRM (Tool Box: http://www.gwptoolbox.org.) The second WWDR "Water a Shared Responsibility" also includes various case studies illustrating progress on IWRM. (See http://www.unesco.org/water/wwap/wwdr/wwdr2).

9.4 China – Provincial Level: Liao River Basin Management

Issues:

• The province of Liaoning with a 41 million population has seen a rapid development resulting in water shortages and severe water pollution. In the 1980s water use efficiency was very low both within urban/industrial areas and irrigation. Water pollution was rampant. No fish could be found in 70 per cent of the streams and ecosystem productive functions had ceased in 60 per cent of the streams. Citizens were ignorant of water conservation issues. Urban wastewater was charged untreated into streams and in some cases infiltrated into the groundwater aquifers. Deforestation took place in the upper parts of the catchments.

IWRM Actions:

- Establishment of an institutional framework comprising Liaoning Cleaner Water Project Office, Liao River Basin Coordination Commission, EU-Liaoning Water Resource Planning Project Office under which an IWRM Planning Project was developed. Under this project water resources assessment was carried out, a reform of the policy for water exploitation and utilization was made, water prices adjusted, a monitoring network established and capacity building within IWRM made. In addition, the cleaner water project was creating wastewater infrastructure, low production/high pollution production was discouraged, pollution prevention and control of Liao River Basin was planned and reforestation was implemented.

Tangible Impacts:

- Reduction of pollution loads by 60 per cent and quality of river water considerably improved. Upstream-downstream conflicts were reduced and deforestation practices halted. Drinking water within the basin was safeguarded and ecosystems in several river stretches were restored. Groundwater pollution was reduced and public awareness of demand management and pollution risks was raised.

Source: EU Liaoning Integrated Environmental Programme, Lagoon, Chief of EU Party Alan Edwards – MWH Environmental Engineering.

9.5 Columbia – Local Level: Conserving La Cocha Lagoon

Issues:

- La Cocha Lagoon is situated in the high Andes in Colombia with the largest wetland system of the Andes. The forests of the basin are exploited for charcoal production being the cause of soil erosion, loss of fertility, faster runoff and greatly reduced biodiversity.

- Another issue is the planned construction of a major dam system to divert water from the Amazonas Basin to the Pacific side of the Andes. Inundation of 3000 ha of grassland and threatening of the livelihoods of local families is among the negative impacts.

IWRM Actions:

- Partnership established between the Network of Private Nature Reserves, Peasants' Development Association and various community organization with facilitation from WWF. Promotion of greater participation in decision-making processes.

- Private forest reserves with sustainable use were encouraged as well as ecotourism. A Lake Defense Committee was formed and plans for establishment of officially protected areas to complement the private reserves were made.

Tangible Impacts:

- Three hundred eighty seven poor families doubled their income and food requirements are met on site. Threats to La Cocha Lagoon and surrounding wetlands were reduced and the Lake Defense Committee worked with Ministry of Environment towards a declaration of the area as a Ramsar Site. The dam system plans were shelved as Ministry of Environment refused an environmental license based on a balancing between downstream benefits and environmental costs.

Source: IWRM Tool Box, Case # 225 – www.gwptool-box.org.

9.6 Morocco – National Level: Management of Scarce Water Resources and Pilots on Pollution Control

Issues:

- Scarce water resources combined with a rapid population increase, urbanization and industrialization makes water a contentious issue with 42 per cent of the rural population lacking access to potable drinking water. Agriculture uses 92 per cent of the country's dwindling water resources. Large variations in water resources in time and space make sustainable management of water resources a key issue. Challenges include the implementation of a water reform decentralizing financial and planning authority for water resources to nine river basin agencies to be created incrementally.

IWRM Actions:

- Improvement of institutions and policies for water resources management following IWRM principles. Best practices in water resources management developed and disseminated. Non-governmental participation in water resources management had increased. Pilots were undertaken among others within wastewater. Actions were undertaken facilitated by USAID.

Tangible Impacts:

- Soussa-Massa River Basin Agency established and operating according to IWRM principles. Multi-agency cooperation and participation of private water user associations in management decisions takes place. National and regional institutional responsibilities have been defined and consolidated.

- Procedures for allocation of water were established together with technical capacities to allocate and monitor water quantity and quality and mechanisms together with technical capacities to allocate and monitor water quantity and quality and mechanisms for communication between sectors and agencies. Pilot projects were undertaken in Fez, Al Attaouia and Draga and included construction of innovative wastewater treatment plants. In Nakhla, watershed soil loss was significantly reduced through soil conservation measures.

Source: USAID Water Team – Case Study in Integrated Water Resources Management USAID/Morocco SO2 Close-Out Report.

9.7 Fergana Valley – International Level: Improving Water Accessibility through IWRM

Issues:

- Once the most fertile valley in Central Asia, Fergana valley with its approx 10 million inhabitants is now subject to high soil salinization and crops no longer suffice to feed the population. State boundaries between Uzbekistan, Kyrgistan and Tajikistan make transboundary management problematic and cause constant internal and interstate disputes. More than 60 per cent of the inhabitants do not have access to safe drinking water and basic sanitation resulting in widespread waterborne diseases in the rural areas. Irrigation infrastructure is inadequate and the water use is inefficient.

IWRM Actions:

- Improved management of water resources based on IWRM principles emphasizing higher efficiency and more equity. IWRM capacity building within river basin management among river commissions, provinces, municipalities, companies and water user-associations. Demonstration of bottom-up approaches increases in yields and water productivity by up to 30 per cent. Swiss Agency for Development and Cooperation assisted the Interstate Commission for Water Coordination in the implementation.

Tangible Impacts:

- Partnership between all water management factors across Fergana valley: Safe drinking water provided to 28 villages with a population of 80,000 people and 320 ecological sanitation toilets have been constructed on a cost-sharing basis. Waterborne diseases have decreased by more than 60 per cent on average and infant mortality has been almost eradicated in all villages despite prevailing poverty. 28 Water Committees have been created to operate and maintain the water systems efficiently with more than 30 per cent participation by women. Expansion of improved-irrigation practices as well as innovative solutions for irrigation canal management and sustainable water-user associations in addition to sustainable financing at canal, water-user association and farm level have been made.

Source: SDC in Central Asia – IWRM. www.swisscoop.uz/en/Home/Regional_Activities/Integrated_Water Resources_Management.

9.8 Sri Lanka – National Level: IWRM and Water Efficiency Plan

Issues:

- Inadequate developed water resources to meet the demands; frequent water-related disasters (floods, droughts etc. associated with climatic changes); low water use efficiency; delay in implementing National Water Resources Policy due to politicization of basic policy issues.

IWRM Actions:

- A baseline assessment of water resources was made under Sri Lanka National Water Development Report (SLNWDR) prepared for WWAP.

- To address the inadequacy of water development, several diversion and storage projects have been initiated. Some were completed recently. A disaster management plan and institutional setup have been implemented too. Sectoral water use efficiency improvement plans are implemented. A National Water Development Report has been prepared under WWAP and it is planned to update this every 3 years.

Tangible Impacts:

- A considerable number of people living in water scarce areas of the country have benefitted through diversions and storage facilities.

- To bridge the water demand/availability gap, several projects are planned and implemented. Ongoing Manic Ganga Project and Weli Oya Diversion Project are nearing completion. Studies on the impact are continuing. In the case of irrigation sector, several irrigation schemes have improved their water productivity. Similar improvements are experienced in drinking water sector. The disaster management institutional setup contributed to mitigate the impacts and provide warning for recent floods. The SLNWDR has created an awareness of water-related challenges among the key stakeholders.

Source: Adapted from WWDR number 2 http://www. unesco.org/water/wwap/wwdr/wwdr2

9.9 USA – State Level: NY City Water Supply as a Partner in Watershed Management

Issues:

- Faced with deteriorating input water quality NY City had the choice of building a new water supply treatment plant at a cost of USD 6,000 million or taking comprehensive measures to improve and protect the quality of the source water in the Croton and Catskill/Delaware watersheds totalling approx. 5000 km² delivering water for over 9 million people in New York City. Dual goals of protecting water quality and preserving economic viability of watershed communities were set out.

IWRM Actions:

- Development of partnerships between NY City, NY State, Environmental Protection Agency, watershed counties, towns and villages environmental and public interest groups. Programmes were developed to balance agriculture, urban and rural wastewater and storm drainage infrastructure, environment and the quality of water in the 19 reservoirs and 3 controlled lakes. A watershed agricultural programme was supplemented by land acquisition, watershed regulations, environmental and economic partnership programmes, wastewater treatment plant upgrades and protection measures at reservoirs.

Tangible Impacts:

*   More than 350 farms within the watershed have embarked on implementation of best management practices reducing pollution loads, acquisition of 280 km² land for protection, enforcement of effective watershed regulations, remediation of 2000 failing septic systems, upgrading of wastewater treatment plants with tertiary treatment. More than 50 per cent reduction in coliform bacteria, total phosphorus and several other major contaminants were achieved. NY City water supply was exempted from filtration, the population of the watersheds enjoys an improved environmental quality and a total saving of USD 4,400 million was realized.

Source: New York City, Department of Environmental Protection, Bureau of Water Supply: "2006 Watershed Protection Programme. Summary and Assessment". www.ci.nyc.ny.us/html/dep/html/ watershed.html

9.10 Kazakhstan – National Level

Management of Scarce Water Resources and Pollution Control

Issues:

*   There are plenty of water-ecological problems serving as obstacle, of which the most acute ones are growing water deficit; Pollution of open and underground waters; Enormous over-norm water losses; Exacerbation of the problem of quality drinking water supply to population; Problems of interstate water apportioning; and Deterioration of the technical state of the dams, waterworks facilities and other installations. Actually, the situation with water management is tense throughout the territory of the republic and the environmental ill-being has overtaken all major river basins of the country.

IWRM Actions:

*   In accordance with the Water Code of the Republic of Kazakhstan, the Water Resource Committee of the Ministry of Agriculture is assigned to manage, regulate the use and to protect the water resources, including renewable water resources. With the purpose of improving the management of water resources and introduction of international practice, the Committee, as of June 2004, has been carrying out the development of Integrated Water Resource and Water Efficiency Management Plan (IWRM).

*   Legal and organizational conditions for transition to integrated water resource management have also been established. Basin Councils – basis for IWRM Plan implementation – have been established to increase the involvement of interested parties in water resources management.

Tangible Impacts:

*   The necessary legal framework, namely after Code, Land Code and Forestry Code (2003), The law "On Sanitary-Epidemic Security of Population" (2003) is established

in Kazakhstan. For the implementation of the IWRM Plan, River Basin Organizations, namely Basin Councils are being created. In the sense of territorial division, the basin councils have been created in 8 hydrographic basins of Kazakhstan as well as in separate water objects.

Source: The Plan of Integrated Management of Water Resources of the Republic of Kazakhstan. A.Y. Nikolayenko and A.K. Kenshimov.

9.11 Mozambique/Zimbabwe – Trans-boundary Level

The Pungwe River Project

Issues:

* During spring tide and low river flows, saline water intrusion extends upstream of Pungwe River mouth, which has a negative impacts on sugar cane farming and domestic water for Beira City in Mozambique. The effects of gold mining activities in the Pungwe basin dominate the water quality and increased sediment concentrations of the surface water of the Pungwe River. The gold mining activities in the river basin are mainly poverty-driven, i.e. it is a subsistence activity. The suspended sediments make the water unsuitable for drinking, washing and irrigation, bury the aquatic fauna, prevent photosynthesis and have effects on the fish population. Miners use mercury in the gold mining process causing elevated concentrations of mercury in the suspended sediments.

* Also other heavy metals, e.g. lead and cadmium are bound to the suspended sediments since they exist naturally in the soils. Floods cause frequent problems in the lower parts of the Pungwe River basin. Widespread poverty and competing demand for available water resources within and between the countries.

IWRM Actions:

* The Pungwe Project commenced in February 2002 and included three phases, viz: Phase 1- Monograph Phase, Phase 2 – Scenario Development Phase, and Phase 3-Joint IWRM Strategy Phase. During the monograph phase a large effort was directed towards improving the knowledge base for the development of the water resources of the basin through several sector studies. The scenarios for water resources development were elaborated in the Phase 2.

* The development scenarios included several projects and studies, including e.g. possibilities of medium large dams on the Pungwe river or its tributaries, flood warning system, local groundwater assessments and measures for improved surface – water quality. In Phase 3 implementation plans for the projects adopted by the stakeholders of the Pungwe River basin were elaborated and the Joint Integrated Water Resources Management Water Strategy formulated. In parallel the development of a climate change adaptation strategy for the basin has commenced. Local assessment of

climate change impacts were made by feeding GCM scenarios into a regional higher resolution climate models and linking it to the hydrological models of the basin.

Tangible Impacts:

*   Sector studies conducted by the project describe the present situation in the basin with regards to water resources, environment and pollution, water demand, infrastructure and socio-economy. River basin organizations have been strengthened, water permitting and revenue collection operationalized and stakeholder participation increased through the establishment of a basin committee. A five year joint programme between the Governments of Mozambique and Zimbabwe has commenced to strengthen the capacity of key basin IWRM institutions – To strengthen and expand stakeholder participation in Integrated Water Resources Management in the Pungwe Basin.

*   To ensure appropriate, efficient, effective and sustainable technical systems and capacities for the collection, monitoring, management and communication of water resources data; To mobilize resources for sustainable, poverty-oriented, water-related development investments in the Pungwe Basin through establishment of a Pungwe Basin Pre-Investment Facility and launching of the Pungwe Basin Initiative. In addition, seven Critical Development Projects have been developed with their own specific objectives. Large-scale investments such as major hydraulic infrastructure is anticipated to be funded through other sources mobilized through a Pungwe Basin Investment Facility. The Joint Integrated Water Resources Management Strategy for the Pungwe River Basin has been able to materialize the vision of a broad and sustainable socio-economic development without environmental harm.

Source: www.pungweriver.net

9.12   Chile – National Level: Impact on Water and Environment due to Macro-economic and Social Development Policies

Issues:

*   Chile's macro-economic growth policies boosted exports, but a sharp rise in demand for water was also evident. Much of this demand occurred in relatively water-poor basins, where it was driven by market forces or the availability of other inputs or resources, and not by the area's water endowments. This has led to growing competition for water in some basins. Policy makers and water planners therefore need to be aware that if economic policies continue to encourage water dependent exports, then ever greater quantities of water will need to be found.

*   Development has placed additional pressure on the environment in general, and on water resources in particular. Over the two decades the use of wells in agriculture has increased six fold, the use of wells for drinking water four fold, and, during the last decade, 40 aquifers have been closed to new concessions.

IWRM Actions:

* Improvements in water-use efficiency have been considerable, especially in those areas linked to exports. Cleaner production practices triggered by globalization have also benefitted the environment. Increased private-sector investment in sanitation has been encouraged by Chile's focus on maintaining its macro-economic equilibrium. This has boosted the development of Chile's sewerage, as well as its water supply sector. New water and environmental laws and regulations have also been put in place. In 2005, reform of the country's Water Code sought to establish a more stable balance between the public interest and the rights of private individuals and among social and productive demands and environmental considerations.

Tangible Impacts:

* Working in water-scarce areas has increased the prices of water rights and forced the mining sector to increase the efficiency of its water use three fold over the last 20 years, while water use in wood pulp production has fallen by 0 per cent ton produced. Macro-economic policies to improve cost recovery have caused household water consumption to fall by 10 per cent, in reaction to a 38 per cent increase in domestic water supply. Some sectors (such as mining, agriculture and wood pulp production) have gone beyond national requirements and agreed to clean production programs accepted globally. The percentage of sewage treated in Chile leapt from 17 per cent in 1997 to 81 per cent in 2005, and by 2010 almost all the country's sewage is likely to be treated.

Source: Water and Sustainable Development: Lessons from Chile, Policy brief prepared by Sandy Williams and Sarah Carriger under the direction of the GWP Technical Committee.

9.13 Uganda – National level: IWRM and Water Efficiency Plan

Issues:

* In the 1990's deteriorating quality and quantity of water resources due to poor land use practices and inadequately regulated use of water and discharge of wastewater, inadequate legal and institutional framework for WRM, reform in the light of decentralization goals. Increased stakeholders' involvement in WRM at both national and local levels is required.

IWRM Actions:

* The National Water Resources Management Strategy is being implemented at both national and local levels. Institutional arrangements at national level involving a 12 member high level Water Policy Committee is being revitalized and the Department of Water Resources Management has been elevated to a Directorate in the Ministry of Water to strengthen the position of water resources management. At local level, catchment management organizations involving a Catchment Advisory Committee, Catchment Secretariat, Stakeholder Forum and Water User Committee are being piloted in one catchment before roll out to a wider part of the country in 2008.

- Strengthening water resources management framework involving water resources assessment and monitoring networks and regulation of use and pollution of water resources through continued implementation of a water permits system.
- Improvement of the enabling legal and institutional framework for WRM at both national and local levels.
- Decentralization of management of water resources to catchment management zones.

Tangible Impacts:
- An enabling legal and institutional framework for WRM is in place;
- Water resources assessment and monitoring networks and a water permits system are fully operational and
- Piloting of decentralization of WRM to catchments is almost complete and roll out to a wider part of the country will be done in 2008.

Source: Adapted from WWDR number 2 http://www. unesco.org/water/wwap/wwdr/wwdr2

9.14 Decision on Water Allocation

- Managers, whether in the government or private sectors, have to make difficult decisions on water allocation. More and more they have to apportion diminishing supplies between ever-increasing demands. Drivers such as demographic and climatic changes further increase the stress on water resources. The traditional fragmented approach is no longer viable and a more holistic approach to water management is essential.

- Countries and regions have very different physical characteristics and are at very different stages in economic and social development, hence there is a need for approaches to be tailored to the individual circumstance of country and local region.

9.15 The Status of UN-Water Report

This Report, compiled by UN-Water, aims to illustrate progress made on meeting the target to develop integrated water resources management and water efficiency plans, with support to developing countries, through actions at all levels agreed at the World Summit on Sustainable Development (WSSD) in Johannesburg, through the Johannesburg Plan of Implementation (JPoI).

The Report is based on a survey covering 104 countries of which 77 are developing or countries in transition and 27 are developed (OECD and EU member states). The survey brings together the results of questionnaires by UN-DESA, and UNEP1 in 2007. Several other members of UN-Water and partner agencies have supported and contributed to the Report. Besides, UNDP, UN Statistics, WHO, WWAP and GWP helped in the preparation of the report. The survey recognizes that countries use different terminology for their water resources management plans.

It provides the most objective and comprehensive overview of the current status of water resources management. The Report also includes information gathered by the more informal surveys conducted earlier by the Global Water Partnership and the African Development Bank.

9.16 Findings

Developed countries: They have advanced on almost all major issues, however, there is still much room for further improvement.

*   Of the 27 countries responding to the UN-Water Survey only 6 claim to have fully implemented national IWRM plans; a further 10 of those countries claim to have plans in place and partially implemented.

*   The Report indicates that developed countries need to improve on public awareness campaigns and on gender mainstreaming.

Developing countries: There has been some recent improvement in the IWRM planning process at national level but much more needs to be done to implement the plans.

*   Of the 53 countries for which comparison was made between the GWP and the UN-Water surveys (conducted approximately 18 months apart), the percentage of countries having plans completed or under implementation has risen from 21 to 38 per cent. On this measure the Americans have improved most from 7 to 43 per cent; the comparable changes for Africa were from 25 to 38 per cent and for Asia from 27 to 33 per cent. However, some of the changes may be due to differences in the questionnaires.

*   Africa usually lags behind Asia and the Americas on most issues; however it is more advanced on stakeholder participation and on subsidies and micro-credit programmes.

*   Asia is more advanced on institutional reform and yet lags behind in institutional coordination.

*   There are many illustrations of the tangible benefits of implementing plans that have adopted the WRM approach. There are examples at the national and international levels; of particular significance are the examples at the community and provincial levels for it is at these levels that so many societal gains can be made.

*   It is clear that many countries consider that plans that follow an IWRM approach automatically also include water efficiency measures. There was considerable ambiguity in the responses concerning water efficiency in large measure reflecting diverse situations. It is recognized that taking actions that make water use more efficient is beneficial for economic and social development and, although many countries indicated through the questionnaires that water efficiency measures were not relevant to their particular circumstances, it should not be implied that such measures should not be considered necessary. It can be concluded from this survey that much more effort needs to be made to incorporate explicitly water efficiency measures within the framework of IWRM.

Development of indicators and Road-mapping Initiative

A great deal of effort has gone into the development of a set of indicators that meet the requirements of being specific, measurable, attainable, relevant, realistic and timely but more work is required.

It is being developed concurrently with this Report and complementary to it is intended to help countries focus on the steps to be taken towards better water management. Drawing inspiration from the IWRM principles and the plans and strategies that they have prepared to help catalyze change. At regional and global levels, the roadmaps could serve as benchmark for monitoring progress in improving water resources management. Indicators and monitoring could provide countries with a better assessment of the needs to advance in their implementation of IWRM.

9.17 Recommendations

The survey indicates that more emphasis is needed in the following areas:

- Countries, particularly those that are lagging behind, need to prioritize the development of IWRM and water efficiency measures, with the help of supporting agencies;
- Countries need to prioritize the implementation of policies and plans once they have been developed;
- Countries should establish roadmaps and financing strategies for the implementation of their plans with External Support Agencies (including the UN, donors and NGOs) providing support to countries, based on demand;
- Experiences in implementing IWRM should be evaluated, monitored and shared through global coordination mechanisms. This will require more work on indicators and follow-up processes that do not add an undue reporting burden on countries;
- The UN World Water Assessment Programme and its associated World Water Development Reports should continue to provide an up-to-date global overview of progress on implementing the IWRM approach.

9.18 Conclusion

There is a recognized need to develop a set of indicators which would characterize the status of implementation of the IWRM approach within countries. There have been many attempts to produce indicators which would adequately encompass diverse situations and the very different time scales at which implementation is taking place. The process is highly complicated and challenging. Moreover, this must be considered in the light of established reporting mechanisms, e.g. Statistics, and avoid adding onerous reporting demands on national governments.

UN-Water has undertaken a major initiative through the World Water Assessment Programme to develop a comprehensive set of indicators – summary of progress is documented in the Second World Water Development Report.

To further develop suitable indicators UN-Water has established a Task Force on indicators, monitoring and reporting. Many indicators already exist to measure social progress and the aim is to add value to these and not reinvent the wheel. A summary of progress made to date by the many agencies and organizations involved has been produced by UNEP-UCC. The road-mapping initiative, being developed concurrently with this Report and complementary to it, lays out a timetable over the next seven years for the development of an achievable set of indicators including those specifically related to IWRM.

Reference

http://www.unwater.org.

Section 3

Water Vision: A Management Paradigm

CHAPTER 10

Improvement in Water Use Efficiency

10.0 Introduction

Water-use efficiency is presently estimated to be 38 to 40 per cent for canal irrigation and about 60 per cent for ground water irrigation schemes. On an average, our country's per capita water availability per year was estimated at 2214 cubic metres against the global average of 9231 cubic metres and 3020 cubic metre, 3962 cubic metre and 4792 cubic metre per year respectively for countries like Afghanistan, Pakistan and Sudan. India was ranked at the 42nd position among 100 countries by per capita water availability.

In the total water use the share of agriculture was 83 per cent, followed by domestic use (4.5 per cent), industrial use (2.7 per cent) and energy (1.85 per cent). The remaining 8 per cent was for other uses including environmental requirements. The projected total water demand by the year 2025 is around 1050 cubic kilometres against the country's utilizable water resources of 1140 cubic kilometres. The share of agriculture in total water demand by the year 2025 would be about 74 to 75 per cent. Thus, almost the entire utilizable water resource of the country would be required to be put to use by the year 2025 A.D. Irrigation, being the major water user, its share in the total demand is bound to decrease from the present 83 to 74 per cent due to more pressing and competing demands from other sectors by 2025 efficiency in general and for irrigation in particular assumes a great significance in perspective water resource planning. It is estimated that a 10 per cent increase in the present level of water use efficiency projects, an additional 14 m.ha area can be brought under irrigation from the existing irrigating capacities which would involve a very moderate investment as compared to the investment that would be required for creating equivalent potential through new schemes. Thus, there is a need to improve the water use efficiency in most of the existing irrigation projects through modernization, renovation and upgradation to realize optimum benefits on the one hand and mitigate the consequential side effects like water logging etc. on the other.

10.1 Improvement in Water Use Efficiency

On a rough basis, it is estimated that with a 10 per cent increase in the present level of water use efficiency in irrigation systems, an additional 14 m.ha area can be brought under irrigation from the existing irrigation capacities at a very moderate investment as compared to the investment that would be required for creating equivalent potential by way of new schemes.

10.2 Improvement in Water Management Systems of Selected Irrigation Schemes

To promote the process of improvement in water management through upgradation of the main systems of selected irrigation schemes the National Water Management Project (NWMP), an externally aided project (EAP), was implemented. The basic objective of the project was to improve the irrigation coverage and agricultural productivity and thereby increase the income of farmers in the command areas through a more reliable, predictable and equitable irrigation service. This project was implemented in states of A.P., Bihar, Gujarat, Haryana, Karnataka, Kerala, M.P., Orissa (now known as Odisha), Tamil Nadu and Uttar Pradesh, covering 114 irrigation projects with a command area of about 3.348 m.ha at a cost of ₹587.81 crore at current prices. The IDA Credit was to finance about 73 per cent of the participating states. With the implementation of this project an overall improvement was found in terms of water management, productivity and farm income etc. Increase in farm income in the 9 schemes on completion as a direct result of the Project ranged from 8 to 89 per cent, the highest of 50 per cent and 89 per cent were in the tail end reaches of the projects which were essentially rainfed prior to NWMP. Although the project outcome has not been rated satisfactory in terms of achievement of the target which was only about 15 per cent completion of command at the time of termination of the scheme, this provided adequate feedback for the formulation of future strategy. Now, the Ministry of Water Resources has initiated follow-up action on NWMP-II with an estimated cost of ₹2880 crore for 7 years.

10.3 Renovation and Modernization of Projects

Increasing the effective irrigation area through timely renovation and modernization of the irrigation and drainage systems, including reclamation of waterlogged and salinized irrigated lands through low-cost techniques, is needed to be considered especially in the context of the present resource constraints. It is estimated that about 21 m.ha of irrigated area from major and medium projects from pre-Independence period and those completed 25 years ago, require renovation/upgradation/restoration to a great extent of the areas which have gone out of irrigation, either partly or fully, due to deterioration in the performance of the systems. The total investment involved is estimated at ₹20,000–30,000 crore over a period of 20 years.

10.4 Externally Aided Projects in Andhra Pradesh

Recently, an externally aided Andhra Pradesh Irrigation Project (Phase-III) has been taken up for modernization/renovation of selected irrigation projects in Andhra Pradesh. Besides the above, Punjab Irrigation and Drainage Projects Phase-II, with an estimated cost of US $165 million including components of all sectors of irrigation (major/medium/minor, CAD and flood control) is also under implementation. This is aimed primarily at better water management and improved functioning to achieve optimum utilization of water in Punjab, as the state has almost exhausted the exploitation of surface water. However, a greater push for modernization/renovation of existing irrigation projects will be needed during the Ninth Plan period.

10.5 Water Resource Consolidation Project (WRCP)

In recent times, the Water Resource Consolidation Project (WRCP) has been taken up in the states of Haryana (estimated cost - ₹1442.12 crore), Orissa (₹1409.00 crore) and Tamil Nadu (₹807 crore) which also envisages, inter alia, the completion of some uncompleted major and medium irrigation projects and strengthening of institutions on the lines of Participatory Irrigation Management/Irrigation Management Transfer (PIM/IMT). Such projects are expected to be taken up in more states during the Ninth Plan period.

10.6 International Cooperation

During the Eighth Plan, the externally aided irrigation projects, put together, accounted for 18 per cent of total estimated cost of ongoing projects. The share of irrigation in the total external aid annually was around 7 per cent to 8 per cent. However, the expenditure incurred on the externally aided major and medium projects during the first three years of the Eighth Plan was only 31 per cent of the target. Thus, vigorous efforts to attract more external investments in irrigation sector as well as to improve the level of utilization are the need of the hour especially against the backdrop of constraints in domestic funding. Apart from provision of inadequate level of funding as per provisions in the agreement, the other major reasons of the low level of utilization of external assistance relate to the tendering procedure and its finalization including insistence on global tender for the material/machinery which may be available in India, participation of NGOs in the land evaluation committees in the states leading to delays in finalizing land acquisition proceedings, frequent review of R&R programme by making field visits and interrogating the oustees about their level of satisfaction which at time encourages them to ask for more and more facilities which, if not fulfilled, results even in stoppage of works and, sometimes, certain techno-economic issues requiring change in the scope of the project during execution although the agreement has been signed on the basis of detailed appraisal earlier etc.

10.7 Irrigation Water Charges

According to the National Water Policy, water rate should be such as to convey its scarcity value to the users and motivate them in favour of efficient water uses, besides, at the same time, being adequate to cover annual maintenance and operation charges and recover a part of the fixed cost. Agricultural productivity per unit of water needs to be progressively increased in order to be able to compete with other higher value uses of water. Simultaneously, sound practices for irrigation revenue recovery by the appropriate agencies, even by institutional adjustments, need to be promoted. Most of the states have at present very low irrigation water rates at substantively varying levels and have not revised these for the last 2–3 decades.

10.8 Revision of Water Rates

A few states had revised the water rates recently but the revised rates in some cases had been withheld by the State Governments. Most of the north-eastern states (except Assam and Manipur) do not even charge any irrigation water rates. Maharashtra is the only state where the irrigation water rates are announced for a 5-years period at a time with provision for yearly escalation so as to cover the full O&M cost as well as the interest payable on the public deposits raised through irrigation bonds. The State Government of Andhra Pradesh, Maharashtra, Haryana and Orissa have revised the water rates recently. The position in respect of minor irrigation including ground water is also not encouraging.

10.9 Suggestion of the Finance Commission

The Tenth Finance Commission have suggested the norms for O&M cost of works at the level of ₹300 per ha. in case of utilized potential and ₹100 per ha. for the unutilized potential with 30 per cent increase of hilly areas and suitable increase of insulating inflation. Accordingly, the estimated total O&M cost per annum for the country would be about ₹2500–3000 crore. Against this requirement, the O&M funds, being provided are actually less than even 1/4th with wide variation from State to State. This is one of the major reasons for the deterioration in the performance in terms of adequacy, timeliness and equity in the provision of irrigation water in the system.

10.10 Setting up of Water Pricing Committee by the Planning Commission

A Water Pricing Committee, an internal group was set up by the then Planning Commission to study the pricing of irrigation water. Some of the salient features of the recommendation of this Committee are:
- Treating water rates as user charge, the objective being ultimately to recover cost.
- Linking revision of water rates with the improvement in the quality of service.
- Revision and implementation of water rates in phases.

- Consolidation of the system of farmer group management.
- Upgrading the system to higher levels of efficiency in water use and productivity.
- Switching over progressively to volumetric water rates structure.
- The setting up of "High Powered" autonomous boards at State level to review the policy establish norms regarding maintenance costs, assess the actual expenditure and determine the parameters and criteria for raising water rates.
- Mandatory review of all matters related to water pricing every five years, etc.

10.11 Recommendations of High Powered Autonomous Board

Subsequently, to go into the recommendations of the above Committee, the then Planning Commission constituted a Group of Officials under the chairmanship of the then Secretary, Planning Commission and members from selected states and concerned Government of India Ministries/Departments. The Group unanimously recommended the following:

- Full O&M cost should be recovered in the phased manner i.e. over a 5-year period starting from 1995–96 taking into account the inflation.
- Subsequently after achieving O&M level the individual states might review the status to decide on appropriate action to enhance the water rates to cover 1 per cent of the capital cost also.
- In addition to the above, the setting up of Irrigation and Water Pricing Boards by all the states and mandatory periodic revision of water rates at least every 5 years with an opportunity for users to present their views were also recommended.
- Further, the Group also recommended the formation of Water Users Associations and the transfer of the maintenance and management of irrigation system to them so that each system may manage its own finances both for O&M and eventually for expansion/improvement of facilities.
- During Ninth Plan, all the states implemented the recommendations of the Group in a first phase of implementing the Water Pricing Committee's Report.

10.11.1 Private Sector Participation

Private sector participation involves not only the private corporate sector but also groups like farmers' organizations, voluntary bodies and the general public. About 90–95 per cent of ground water development is by private efforts either through own financing or institutional financing or both. However in the case of surface water, especially major and medium projects, all the irrigation projects are not equally endowed with the potential for privatization and, as such, identification of projects as a whole or partially (i.e. planning and investigation, construction, operation and management financing and maintenance etc.) may have to be undertaken in the light of its viability vis-à-vis various privatization options are available with hydel power generation and recreation, etc. along with irrigation, the viability for privatization of a project improves.

10.11.2 Actions Initiated by Some States

Some states like Maharashtra, Madhya Pradesh and Andhra Pradesh have initiated the action for privatization of irrigation projects. These projects are envisaged for privatization on Build-Own-Operate (BOO), or Build-Own-Operate-Transfer (BOOT) or Build-Own-Lease (BOL) basis. In the case of projects on BOO basis, the Irrigation Department may buy water in bulk from the agency at mutually agreed price for distribution to the farmers. Apart from this, Maharashtra Krishna Valley Development Corporation (MKVDC) for Krishna Valley Projects, Sardar Sarovar Narmada Nirman Ltd. (SSNNL) for Sardar Sarovar Project in Gujarat and Jal Bhagya Nigam for Upper Krishna Project, Karnataka have mobilized financial resources through issue of Public Bonds from the private market.

The deliberations in the Workshop indicate that private sector participation in irrigation and multipurpose projects is feasible but selectively. Some procedural and legal changes are required to be undertaken in respect of clearances of projects and involvement of private sector investors in this regard. More specifically, some suggestions as indicated below in brief have been offered by the participants:

- Private sector participation could be thought of on BOL or BOLT basis for a specified period of say 10–30 years.
- While it may be more suitable for medium and minor projects, it could pose some problems in the case of major projects.
- Clearances such as forests, environment, resettlement and rehabilitation, acquisition of land, etc., should be carried out by the Government departments.
- Concessions should be offered to private sector investors to augment their revenue. These may include tourism, water sports, navigation, moratorium on loans, tax concessions, etc.
- Distribution of water after bulk supply to Water Users' Associations (WUAs) should not be handled by private sector. The WUAs should be encouraged to be formed and they should manage distribution.
- Safety and sociological aspects should be looked into the Government departments.
- There should be a guarantee on the return of investment of the private sector.
- In difficult terrains, there should be investment from the Government side also.
- While broad national policy guidelines on private sector participation may be framed by the Centre, details may be worked out by the states as suited to their conditions within the framework of such policy and guidelines.
- The obligations of the Government departments and the private sector should be clearly spelt out in the agreement for such participation. It should also include penalty clauses applicable to both the parties so that slippages do not occur in implementation.

10.12 Conclusion

On the basis of views expressed by the various states, in general, it has been emerged that improvement in agricultural productivity from irrigated agriculture is one of the main objectives of the CAD Programme. An analysis of time series data on productivity in respect of selected projects under CAD Programme indicated, among others that staple crops like paddy and wheat have registered an increase in productivity by 50 per cent (Pench, Maharashtra) and 85 per cent (Gurgaon, Haryana) respectively. Experience has shown that in most of the commands, the main problems are lack of a single window delivery system, poor maintenance and water management of micro-networks, weak extension service for agricultural needs and near-absence of farmers' participation, etc. As such, in most of the CAD projects, the implementation has been limited mainly to the construction of field channels.

CHAPTER 11

Water Crisis: Agenda for the Country

11.0 Introduction

Earth's 97 per cent water, which is in the sea, is very expensive to desalinate. Only about one per cent water is readily available for use and the irrigated agriculture consumes over 70 per cent of water used by people. Industry accounts for more than 20 per cent. Households use only about 8 per cent. In other words, the domestic water consumption is the only use that is practically minimum to assure adequate health, people need a minimum of about 1000 litres of water per day for drinking, cooking and washing. In industrialized countries people are using as much as 450 litres per day, while in the developing countries consumption is as low as 20 litres per day. In water scarce places, the water consumption is not even 20 litres a day. The fact is that more than 26 countries in the main water deficient regions of Africa, West Asia, South Asia, North America, South America, and Australia have already been experiencing water scarcity. And, according to the estimates, over three billion people of about 65 countries will be affected by the year 2025. India will be one among the water scarce countries. The most tragic part is that water is a finite natural resource and no technology, till day, has successfully contributed to making or producing water for mass consumption. In the context of India, the poverty-stricken regions tend to be in the climatic zones subject to drought and other water problems as they are least able to afford alternative sourcing. The crisis about water resources development and management thus arises because most of the water is not available for use and secondly it is characterized by its highly uneven spatial distribution. Environmentalists like Sunderlal Bahuguna are not wrong in saying that the acute scarcity of water may force the powerful nations to wage a new global war for the control of the depleted sources of water. Such a parallel makes the future picture essentially depressing.

11.1 India – A Victim of Uneven Spatial Distribution of Water

India is a victim of uneven spatial distribution of water. For example, the hilly state of Uttaranchal houses several prominent water bodies but thousands of villages in the state still have to do with insufficient water availability. Similarly, despite being a part of the Ganga river basin, Rajasthan suffers from acute water shortage. Therefore, the importance of water has to be recognized and greater emphasis laid on its economic use and better management – simply because in our country, inefficient utilization of water has already created a critical situation.

Water is scarce even for drinking purpose, not to talk about its availability for agriculture, industries, etc. India already faces an alarming situation. Its fragile water resources are stressed and depleting while various sectoral demands are growing rapidly even as about 200 million people in the country do not have access to safe drinking water and nearly 1.5 million children under five die each year due to water-borne diseases. The drought conditions in several parts of the country like Gujarat, Rajasthan, Orissa and Andhra Pradesh are unfortunately on rise. And, disputes over sharing of the water resources are becoming grimmer.

11.2 Improper Management of Available Fresh Water Resources

India will face a severe water crisis if the available fresh water resources are not managed properly. According to the estimates, by the year 2025, the country will face a severe water shortage leading to serious struggles. Also, it is feared that within a few decades the availability of water in the country will be about 1700 to 2000 cubic metres per person as against the world average of 5000 to 9000 cubic metres per person.

11.3 Imbalance between Demand and Supply of Water

In one-third of India's agro-climatic regions, there is water scarcity already in terms of per capita demand and supply of water. This imbalance is bound to lead to conflicts at the local, state and the national levels. At present thousands of Indian cities do not have sources of water and in future: it would have to be transported over larger distances as the water sources move much more away from the cities. The country's current and future situation can be gauged by the trend in water availability. Presently, six of India's 20 major river basins already fall into water scarce category. By the year 2025, five more river basins are feared to be water scarce. Even Brahmaputra, Barak and west flowing rivers will be water insufficient in the times to come. The glaciers of Himalaya are rapidly melting and it is feared that within next 50 to 80 years, most of the glaciers will disappear. The Gangotri glacier has already shrunk backwardly by several hundred metres. In such a scenario, there will be considerable changes in the runoff pattern of the rivers like Ganga causing floods, loss of property and life, and loss of agro-production.

11.4 Water Management an Innovative Course for the 21st Century

For the 21st century, water management must take an innovative course – a course that recognizes water as both a basic need and as a scarce resource. It is very simple to know that if there is scarcity of water, the ability to develop economically is limited. Therefore, three major issues like availability, quality and access must be addressed. Water and sanitation improvements could reduce child mortality by more than one-half. The fundamental relationship between water and survival has long been recognized. Now, it is time to ensure that our finite supply of this valuable resource is used optimally. Compromise will put the future of the country in danger.

11.5 "Water a Scarce Resource to be Managed and Protected" – Message of the United Nations

The United Nations' message is loud and clear: "Water is a scarce resource to be managed and protected". India is not like Canada, a country blessed with 9 per cent of the world's renewable water supply. India has to remember that a large population in the country live with severe water problems many of them, without clean drinking water and sanitation facilities.

One had not heard of river water disputes till about the middle of the 20th century. Disputes and riots over river water are of a recent origin. And quite possibly, rivers and the distribution of their waters will become one of the most politicized ecological issues in the near future. In India, we are already in the grip of water disputes. The dispute over distribution of Cauvery water is already a bone of contention between Karnataka and Tamil Nadu. Similarly, claims of Uttar Pradesh (UP), Haryana and Delhi on Yamuna water also cause problems from time to time. And, now the latest light is emerging over sharing of water of the rivers originating from Uttaranchal and passing through UP. In short, the co-modification of river waters has to be reversed, no matter at what cost.

11.6 Regulation and Development of Water of Rivers

As most of the rivers in the country are inter-state, the regulation and development of water of these rivers is a source of inter-state differences and disputes. However, Parliament has power to make laws with respect to any matter for any part of the country notwithstanding such a matter is enumerated in the State List. In case of disputes relating to waters, according to Article 262, Parliament may be law provide for the adjudication of any dispute or complaint with respect to the use, distribution or control of the waters of, or in, any inter-state river or river valley. Only thing is that this has to be applied judiciously, justly and without parochial approach. Governments' dilemma can only make the things more serious.

11.7 Global Drinking Water Supply and Sanitation Decade

The 1980s was the global drinking water supply and sanitation decade. And during this period, significant advances in water management practices were made, but the goal of universal coverage remains unrealized. Therefore, the task of providing water and sanitation to all and protecting the world's water is urgent as are the fundamental issues related to the access and shared usage. Successful water management will be contingent on integrated strategies, which address many considerations. Sustainable solutions could have economic, environmental, social, political or health implications. Technical expertise in the areas of sanitation, drainage, identification of sources and waste management is critical. Initiatives should involve the use of appropriate technologies and institutional strengthening at all levels. Also, components like health protection, social mobilization, community development, empowerment and capacity building for communities and the participation of the private sector has to be incorporated vigorously.

11.8 India's Water Policy

India's national water policy aims at planning, developing and conserving the scarce and precious water resources on an integrated and environmentally sound basis keeping in view the needs of the state governments. The policy envisages strategies; inter alia, ground water development, water allocation priorities, drinking water, irrigation, water quality, water zoning, and conservation of water, flood control and management. The state governments make their water policies within the overall framework of the National Water Policy.

If people think that the 2,525 km long Gangetic lifeline, whetting the life and appetites of nearly 100 big and small towns along it, will run its revered course, all the while nourishing 40 per cent of the Indian population forever, they are wrong. We now see decline in the number of water projects. Tehri is already in question due to protests by the environmentalists and the unprecedented slow work. The higher project costs and increased competition for funding has made things from bad to worse.

11.9 Scarcity of New Sources of Water

New sources of water are becoming increasingly scarce, more expensive to develop, and more dependent on expertise and technological know-how for planning, design and implementation. Many donors reviewing their participation in this sector causing increasing delays in the implementation of projects. The situation, therefore calls for an urgent soul-searching. With millions of wells scattered throughout rural India and entrenched in traditions of private ownership, protection of drinking water sources, groundwater recharge, and environmental concerns such as water quality problem are the key issues needing to be addressed. Also, pollution or deterioration in water quality can reduce the availability of water in ways that are far less reversible.

11.10 Broad Approaches

Besides, broad approaches are needed to monitor and address environmental impacts and concerns such as water-logging and pollution. These need to be integrated effectively into groundwater development and management approaches. It is regrettable that the attempts to regulate groundwater through restrictions have had only limited success. There is a need for wider dissemination of access to water-related information among people. The effective management of groundwater utilization requires strong data base and analytical input like descriptions of the groundwater availability and the functioning of hydrologic system. Therefore, the groundwater legislation is essential for management. The interim order issued by the Supreme Court of India establishing the Central Groundwater Board as a central groundwater authority provides an important opportunity for developing and passing effective legislation and corresponding regulatory and management mechanisms at the Central and State levels. Subsidized electricity in the rural areas will have to be discouraged in this regard so that people do not resort to maximum pumping groundwater.

11.11   The World Bank Assisted Swajal Pariyojana for the Water and Sanitation Programme

The World Bank assisted five-year Swajal Pariyojana for the water and sanitation programme was started in 1996 in the water scarce regions of Uttarakhand and Bundelkhand of the then Uttar Pradesh. The scheme envisaged covering more than 900 villages in Uttarakhand and Bundelkhand. It is unfortunate that the Swajal as a community-based programme is an absolute failure as the village water and sanitation committees formed to look after the planning, execution/implementation, monitoring and maintenance of the project concerned, were forced to become redundant due to undue interference of the state officials. Therefore, the efforts of the bodies like the World Bank to commodify water must be discouraged. Besides, the traditional knowledge has been ignored while implementing the scheme. Cannot sponsors of the projects like Swajal take a lesson from Rajendra Singh, the Waterman of Rajasthan.

11.12   Indigenous Knowledge

The history of water harvesting in the country is very old, dating back to several thousand years, the mode being collection and storage of rain water, run-off water and water from the flooded rivers. Today, efforts are being made in several parts of the country in this direction. Even, the judiciary has intervened in the matter. It has been made mandatory in the city like Delhi that for houses beyond a specified area limit are being constructed after certain period, making arrangements for in-house water harvesting is necessary.

11.13   Social Mobilization – Most Important

But, what is most important is social mobilization. Mere directives are never effective on a long-term basis. Any such activity must have the community involvement. Rajasthan's waterman and the Magsaysay Award winner, Rajendra Singh, has had very bad experience with the government agencies. When he, along with his colleagues, began their first project at Gopalpura village, the government proved to be a hindrance. After they turned the river Arvari into a perennial rives through water harvesting, the government started to give out licenses for fishing in that river. M/s Singh had to fight tooth and nail to save the river from the clutches of fish contractors. His water management movement, apart from bringing agricultural prosperity to the region, has also helped in raising the cattle population. According to Mr Singh, the forests cover has also increased in the Sariska Tiger Reserve.

11.14 Indigenous-folk System of Knowledge and Exogenous-Scientific Knowledge Bases

Another price for new technologies is the emerging tension between indigenous-folk system of knowledge and exogenous-scientific knowledge bases. Traditional subsistence methods are based on bodies of knowledge that have evolved through trial and error over the centuries and are highly adaptive to the constraints of specific highland and are sustainable without long-term damage to the land. In addition, these methods are not dependent on alternative market-based resources.

11.15 Erosion of Local Knowledge Affects Households'

The erosion of local knowledge affects households' ability to adjust to emergencies and, in many instances, also leads to the devaluation of women who are the main repositories of this knowledge. Therefore, the role of local people in the planning and development of water programmes in their own communities is essential. They offer considerable local knowledge and know-how, and the likelihood of their on-going involvement and ownership results in susceptibility. Particularly, the involvement of women in the water sector leaps logically from their traditional roles as users, providers and managers in terms of household hygiene. Women, and to a lesser degree children, generally obtain, transport, store and then use water in the home.

11.16 Integrated Management Approach to Conserve Natural Resources

Several Indian experiences have hit success. Ralegan Siddhi, a village in Maharashtra's Ahmednagar district, is just an example, where people turned the semi-arid village into a village with no water scarcity, thanks to the efforts of Anna Hazare, an eminent social worker. Once drought-prone, the residents of Ralegan Siddhi have water round the clock. The Planning Commission had proposed in 2000 an integrated management approach to conserve natural resources like water, soil, biodiversity and forests, beside a well-organized monitoring and evaluation process. This was proposed with the aim to bring the natural resources under one umbrella and for the adoption of an integrated management approach. Unfortunately, each of the natural resources – water, soil, forests and biodiversity has been dealt with in a fragmented manner by different agencies.

11.17 Information-Based Efforts

Information-based efforts need to be initiated by the communities by empowering themselves at the grassroots levels. India should give a lead to the recognition that water is a scarce commodity. Efficiency in water use and development of water should be done hand in hand. Also, there is an urgent need to change the mindset of the people and institutional framework to face the water crisis looming over the country. It is not clear what has been done to the proposal of the Planning Commission that was aimed at equity and economic viability. Also, it is essential to devise a set of criteria and indicators (C&I), which would be applied by all those responsible for integrated management of soil for sustainable management of natural resources, including water.

11.18 Clarity in the Criteria Applied to Monitor and Evaluate the Natural Resources

Besides, there is an urgent need for clarity in the criteria applied to monitor and evaluate the natural resources as well as yield regulation. For increased flow of water, the goal of 33 per cent forest cover in the country has to be achieved combined both agricultural and forest policies. And, there has to be a sound system of substitution worked out to ensure forests are not over-exploited and biodiversity is not damaged yet sustained production took place to meet the economic needs of the local people.

The CPCB has identified as many as 22 rivers as polluted in the country and there is no doubt that the rivers have a natural capacity to cleanse themselves but with the growing urbanization, agricultural demand for water increasing and sewage spewing into our depleted river system, this innate capacity to rejuvenate is being incapacitated. Yamuna and Sabarmati are the most polluted rivers of the country. Pollution levels rise phenomenally when the water in the rivers decreases.

11.19 Disturbing Reports

There are also disturbing reports of Ganga drying up because the Gangotri glacier, its main source, is receding at the rate of 10 to 30 metres a year. Most rivers are facing a water shortage heightening the pollution level. In the last 20 years, the area under agriculture has been augmented with increased irrigation drawn from our rivers. Ganga, which runs through one of the most densely populated areas in the world and home to nearly 400 million people, is also heavily polluted. As access to sewer and sanitation facilities in the river basin is so scarce, dozens of cities spew millions of gallons of untreated human and industrial waste into its sluggish waters. Once one could see river dolphins in Ganga near Varanasi city, now even fish are few. Sad to say, all rivers in India are grossly over-exploited and heavily polluted.

11.20 Regrettable Environmental Failures of the Ganga Action Plan (GAP)

One of the most regrettable environmental failures is the Ganga Action Plan (GAP), which was launched in 1985 at a cost of ₹60 crores. A study found that the amount of sewage flowing into the Ganga has doubled since 1985, while a government audit found that funds meant for cleaning the river were being siphoned off. It is now recognized that the major shortcomings of GAP-I. Even the government had admitted its failure in Parliament. GAP-I has already been declared closed and work under GAP-II is likely to be "completed" by 2005. When the total cost of environmental degradation is considered, it more than offsets the positive economic growth of the past two decades, according to the National Environmental Engineering Research Institute.

11.21 Low Compliance of Stringent Pollution Control Steps-Weak Political Will

Although the nation has stringent pollution control steps, compliance is low due to weak political will, outdated technologies, lack of capital and poor infrastructure, so activists say. Saifuddin Soz as Minister for Environment and Forests had to lament over the lethargy and corruption that had ruined a dream project. Under the GAP, sewage treatment plant (STPs) were to be put in 27 Class-1 cities. By the time they were constructed the amount of sewage generated by the city had doubled in some places and the plants just could not handle the load. In the Tenth Five Year Plan a sum of ₹3, 000 crore has been earmarked for cleaning up rivers under the National River Action Plan. The plan will focus on 149 cities along these rivers.

11.22 Scenario

In fact, the water pollution scenario is quite frightening. With the population explosion, urban centres are spreading and there is greater generation of wastewater. 16,000 mld (million litres daily) is generated from Class-1 cities. Delhi alone generates 2,250 mld of sewage, which is more than that of all the Class-2 cities. The low level of the Yamuna and the huge quantity of waste it receives have given it the dubious distinction of being the most polluted river of the country. And of the 17,600 million litres of waste water generated in the country every day, only 4,000 million litres are treated. Vast amount of untreated wastewater is getting into our water bodies and the environment.

11.23 Industrial Effluents and Domestic Wastes Flow into River

Of the 456,000 km length of our rivers, 6,000 km have a bio-oxygen demand (BOD) above 3 mg/l (milligrams per litres), which means they are unfit for drinking according to the CPCB. Yamuna has a BOD 35 to 40 mg/l. The coliform content in Yamuna is as high as they are in raw sewage, according to the National River Conservation Authority in the Ministry of Environment and Forests.

According to several environmentalists, most of India's rivers are already dead, because millions of tons of industrial effluents and domestic wastes flow into them daily. According to Mr M.C. Mehta, environmentalists, among the nation's 18 principal waterways, most are already dead due to the dumping of untreated waste. Who knows Delhi will not be a desert if Yamuna vanishes in the city by 2025.

11.24 Improvements in Water Facilities

The improvements in water facilities can offer both health and socio-economic benefits, especially, for women as closer water supplies can reduce time, energy and physical strain, allowing them to redirect their labour to other important development tasks. Similarly, improvements in the health and well being of children indicate progress. A vibrant water quality monitoring in India is needed desperately as contamination of groundwater and lakes is already devastating the ecosystems. While technological improvements are helpful, the constraints (political, institutional and social) are the crux of the problem. Therefore, an integrated, interdisciplinary approach to water management is necessary.

11.25 Policy Instruments

Policy instruments such as water pricing and cost recovery can be potential contributors to conservation. The users can use these tools to ensure reliable, equitable, rational optimization. Water management will mean better conservation of water supplies by reducing and reusing. For the future, the fundamental approach and the mantra should be: the management of water as a finite and crucial resource.

The order issued by the Supreme Court establishing the Central Groundwater Board as a Central Groundwater Authority provides an important opportunity for developing and passing effective legislation and corresponding regulatory and management mechanisms at the central and state levels. The report also points out that the subsidized electricity in the rural areas encourages individuals who own wells to maximize pumping of groundwater and sales to neighbouring farmers in informal water market.

11.26 Conclusion

India is considered one of the vanguards of environmental protection, a country that is committed to the elimination of environmentally harmful processes and over-exploitation of non-renewable resources. She even created a separate department of environment in 1980 and subsequently upgraded it to a full-fledged Ministry of Environment and Forests in 1985 with the aim to plan, promote and coordinate the environmental and forestry programmes.

However, the results have not been as enthusiastic as expected simply because no serious efforts were made to transfer the required technology to the developing societies living in the various regions of the country. The Rio Declaration on environment and development adopted at UNCED in 1992 remains to be fulfiled. The declaration had stated: "Eradicating poverty and reducing disparities in living standards in different parts of the world are essential to achieve sustainable development and meet the needs of majority of people." Therefore, we have no option but to depend on the available sources of water and have the responsibility to save, conserve and strengthen them without fail. Their conservation and creating conditions (like forestation) for new sources are the only answer simply because as long as there is life on earth, demand for water is going to increase.

CHAPTER 12

Financial Institutions and Water Resources

12.0 Introduction

One hypothesis is that World War III will be fought not over oil resources but over scarce natural resources in general and over water, in particular! There has been a considerable consciousness in India and internationally too about water in recent years and the formation of the National Commission on Integrated Water Resource Development and the Report of the National Commission on Water, are recent manifestation. With a population of 1.30 billion people in 2011, India has 16 per cent of the world's population, 2.5 per cent of the world's land resources and 4 per cent of its water resources.

While the average annual precipitation over India by rain and snow is around 4000 cubic kms, the run-off estimated in our rivers is 1953 cubic kms. The National Water Commission has estimated that the annual usable water resources of the country are 690 cubic kms of surface water and 396 cubic kms of groundwater making a total of 1086 cubic kms. The present quantum of use is estimated at around 600 cubic kms. Thus the position of the water requirement is estimated to vary between 973 and 1180 cubic kms (under low and high demand conditions) by 2050 and that means demand would affect the peninsular rivers which are monsoon-fed while the snow-fed northern rivers could also be affected. Thus, there is a need to utilize the waters available to the country as water is a finite quantity and water is neither added nor destroyed. The major factor is pollution due to urbanization/development which renders water unfit for consumption or irrigation and which could devastate the entire water availability scenario.

12.1 Role of Development Financial Institutions regarding Availability of Water

Development Financial Institutions have major role to play, especially in rural areas, about water availability and their objectives are:

1. Ensure access to safe drinking water as part of Rural Infrastructure Development Fund (RIDF).

2. Ensure adequate availability of water for agriculture and industry.
3. Find appropriate solutions for drought-prone areas and arid zones.
4. Protect and preserve natural environment/ecological systems as part of watershed development or groundwater development programmes.
5. Ensure major inter-state projects involving water storage and management.
6. Assist people to cope with floods/droughts and minimize damage.
7. Ensure more irrigation for enhancing food crop yields.
8. Ensure more traditional community-managed water-harvesting systems.

Development Financial Institutions need to take a more proactive role in the above, as far as water is concerned. A detailed analysis of Development Financial Institutions role in these areas is need so as to ensure a stake in all future developments.

12.2 Drinking Water

The National Water Policy assigns highest priority to safe drinking water but even after 50 years of independence, the number of villages without safe drinking water is growing larger despite annual targets being achieved. The burden of bringing in drinking water falls on women and girl children in rural areas and in urban areas, most large cities are chronically short of water, e.g., Delhi, Mumbai, Chennai, Hyderabad, Jaipur, Nagpur, etc. The recent survey revealed that 30 per cent of revenue villages and hamlets have problem of drinking water. NABARD has taken up some RIDF projects in certain states. We will have to ensure technical solution that there are no slip-backs!

12.3 Water Harvesting

India's ancient tradition of community-based water harvesting systems is declining and a state-managed water system causes canal wastage and pollution. The tanks in Southern India are a prime example of how traditional systems have been discarded to the detriment of the common people. Water harvesting combined with village ecosystem management, as in Ralegaon Siddhi (Annasaheb Hazare's village) and other areas, could start a chain of highly synergistic and substantial ecological and economic changes for the rural villages. Water harvesting and water management are not meant only for watershed areas taken up for treatment but need to be inculcated in all of India's villages.

The components of such a widespread programme are as follows:
1. Water literacy campaign to ensure people's participation.
2. Involve panchayats and officials dealing with rural and urban development in strengthening water harvesting components of these programmes.
3. Consider appropriate technologies including ancient/traditional community-based water harvesting technologies which have stood the test of time.
4. Consider institutional mechanisms needed to promote water harvesting including credit linkages and NGO participation.

Rainwater harvesting ensures that there is production by the masses and not mass-production alone. The Kundi system in Rajasthan's deserts, the canal system in Bihar and Bengal, the Zabo system in Nagaland, the tank systems of Southern India are all important indigenous rainwater harvesting technologies. Development Financial Institutions have done little to finance such schemes. Even tank repairs schemes with refinance link-up have been rare (Andhra Pradesh and Tamil Nadu). Development Financial Institutions need to take a more proactive role in popularizing traditional rainwater harvesting through NGOs and the State Governments.

12.4 Benefits of Irrigation

Canal irrigation benefits are evident in all irrigated areas but efficiency (about 35–40 per cent) is below international standards. The wastage of water resources due to subsidies, needs to be replaced by a regime of relentless pursuit of ensuring maximum value of crop per unit of water. The growth of sugarcane in Maharashtra is a prime example of the failure to ensure more crops per drop of water. The yields of irrigated agriculture in India have been very modest when compared to those of China and other countries. Availability of free canal water, over-application of water, inadequate drainage and failure to take the groundwater table into account, has resulted in water logging (2.46 m.ha) and salinity (3.30 m.ha) in many areas, resulting in valuable agricultural land going out of use. Canal irrigation has led to inequalities with head-reach farmers going in for water-intensive crops while tail-reach farmers suffer from inadequate water, ensuring a water-use pattern that cannot ever be changed. Increasing affluence of large farmers ensures that their economic affluence is converted into political power with the potential for influencing policy-formulation and planning, design and location of further irrigation projects as also their operations, to their advantage.

12.5 Drought-prone/Arid Zones

States like Rajasthan, Gujarat, Maharashtra, Orissa, Tamil Nadu, Madhya Pradesh and Andhra Pradesh have been drought-prone for centuries, with droughts as a recurring feature, causing misery to human beings and cattle, resulting in large-scale devastation and consequent migration. The causes of the drought of 2000 have been analysed and experts opine that the droughts were due to bad water management, due to failure in harvesting rainwater, failure to recharge aquifers and excessive groundwater extraction and consequent water table falling sharply over the years, so that there were no groundwater resources to fall back upon. Yet villages which had ensured better water harvesting and management continued to flourish and to use available groundwater.

12.6 Micro-Water Management

Micro-water management is the key to end drought but then, Development Financial Institutions policies are the same for all areas. Development Financial Institutions needs to create different schemes for different areas with different technical and capital parameters, to ensure micro-management of watershed areas. This needs a very, very flexible approach for which the rural credit agencies like commercial banks, cooperative banks and RRBs should develop competencies and capabilities to handle. This is one of the biggest challenges facing the country, year after year and precious little has been done on the credit front to tackle these issues.

12.7 Flood Control

Another problem is the efforts to "control" floods through embankments and structural means which are costly and often fail due to technical constraints and poor construction causing greater damage down-stream. Also, there is no disaster planning in flood-prone areas leading to excessive damage to rural assets and human misery. Proper conservation of green lands and forests, ensuring that precious top-soil is not washed away, is a much cheaper alternative but which has not been tried out. The continuing destruction of the forests and green lands ensures that floods continue, year after year, in the Brahmaputra Valley in Assam, Central Orissa, etc. Development Financial Institutions could evolve area-wise schemes for ensuring flood water control and drainage.

12.8 Groundwater Misuse

There has been a continuous misuse of scarce resources to ensure groundwater exploitation which is actually harmful to the farm-lands. Over-extraction has led to the groundwater table depletion and salinity is in progress in coastal zones such as Gujarat, etc., with water-logging and salinity in areas of Uttar Pradesh and Bihar. Cheap electricity for farmers has also ensured over-exploitation of groundwater resources and emergence of groundwater markets through tubewells and borewells. This leads to unsustainable extraction of groundwater and inequitable relations between buyers and sellers.

12.9 What Can Development Financial Institutions Do?

Clearly, what has been stated above is a very depressing scenario and the continuance of such unsustainable schemes will lead us into disaster. We need the water requirements on a district-to-district basis as well as on a micro-watershed basis. The local communities must be involved through the efforts of NGOs and must plan their future more effectively. Technical solutions as also credit applications must match the ecological needs on a sustainable basis. Our District Development Managers (DDMs) could be used to prepare such micro-watershed plans and evolve appropriate technologies.

We need to review our unsustainable groundwater exploitation programmes which have been extensively used and concentrate instead on rainwater harvesting which our ancestors used so effectively to create a rich and viable India since ancient times until the Industrial Age when western nations overtook us due to better technologies. Rainwater harvesting would be a cheaper alternative and would not create environmental disasters.

A massive boost must be given to our watershed development programmes in all states. We need to work with institutions like CSE, New Delhi and MIDS, Chennai, for appropriate water-based technologies and development schemes.

12.10 Conclusion

Water is our natural capital and lies at the heart of the economic capital in rural areas. Rural economy consisting of agriculture, animal production, trees and forests is designed entirely around water availability. Water brings the land to life and yields biomass in the form of food, fuel, manure, timber and milk. Bringing water to India's villages is infusing a new economic life into our poverty-stricken rural areas and moving India towards poverty eradication. The way we change our approach from water exploitation to water conservation, would enable us to rebuild India's society at its very roots. The entire community has to participate in these water-harvesting efforts as water structures can be developed, sustained and maintained only by the community as a whole and enable sharing of water resources.

CHAPTER 13

Environmental Sustainability and Protection of Natural World

13.0 Definition of Environmental Sustainability

Environmental sustainability involves making decisions and taking action that are in the interests of protecting the natural world, with emphasis on preserving the capability of the environment to support human life. It is an important topic now, as people are realizing the full impact that businesses and individuals can have on the environment. Environmental sustainability is about making responsible decisions that will reduce our business' negative

impact on the environment. It is not simply about reducing the amount of waste we produce or using less energy, but is concerned with developing processes that will lead to businesses becoming completely sustainable in the future.

13.1 Environmental Sustainability a Topical Issue

Currently, environmental sustainability is a topical issue that receives plenty of attention from the media and from different governmental departments. This is a result of the amount of research going into assessing the impact that human activity can have on the environment. Although the long-term implications of this serious issue are not yet fully understood, it is generally agreed that the risk is high enough to merit an immediate response. Businesses are expected to lead in the area of environmental sustainability as they are considered to be the biggest contributors and are also in a position where they can make a significant difference.

13.2 Some of the Common Environmental Concerns

Businesses can potentially cause damage to all areas of the environment. Some of the common environmental concerns include:
- Damaging rainforests and woodlands through logging and agricultural clearing.
- Polluting and over-fishing of oceans, rivers and lakes.
- Polluting the atmosphere through the burning of fossil fuels.
- Damaging prime agricultural and cultivated land through the use of unsustainable farming practices.

13.3 Many Large and Small Organizations Guilty of Polluting the Environment

For much of the past, most businesses have acted with little regard or concern for the negative impact they have on the environment. Many large and small organizations are guilty of significantly polluting the environment and engaging in practices that are simply not sustainable. However, there are now an increasing number of businesses that are committed to reducing their damaging impact and even working towards having a positive influence on environmental sustainability.

13.4 Environmental Sustainability and Business

Environmental sustainability forces businesses to look beyond making short-term gains and look at the long-term impact they are having on the natural world. You need to consider not only the immediate impact your actions have on the environment, but the long-term implications as well. For example, when manufacturing a product, you need to look at the environmental impact of the products entire lifecycle, from development to disposal before finalizing your designs.

13.5 Business Case for Environmentally Sustainable Practices

Establishing a business case is an important step in assessing the viability of environmentally sustainable practices. If there was no business case behind operating an organization in an environmentally sustainable way, then it simply would not be practical to expect businesses to consider it as an option.

Fortunately, there are many points that strongly support the business case for environmental sustainability. The first point to consider is the fact that moving towards environmentally sustainable practices presents few or no risks to business operations. If a business acts now and environmental sustainability continues to become an increasingly important and heavily regulated issue (as it is likely to do), one will have a head start over many of your competitors. Besides some initial outlay involved in moving towards environmental sustainability, there are not likely to be any long-term negative impacts or expenses incurred.

13.6 Heavily Regulation on Environmental Sustainability Costs Significantly

However, if one fails to act now and environmental sustainability becomes more heavily regulated by governments, there could be significant costs involved. For example, regulators may begin charging businesses based on their negative impact on the environment, leaving one to play catch up and incurring expenses in the process. There may also be incentives schemes introduced that will benefit businesses that operate at better than the minimum standards in relation to environmental sustainability, providing businesses that are sustainable with a clear advantage.

13.7 Environmental Sustainability Reduce Business Expenses in the Medium to Long Term

Another key point in the business case for environmental sustainability is the potential to reduce one's expenses in the medium to long-term. For example, making one's business more energy efficient will save one a significant amount on energy costs and help one to improve one's bottom line. Performing a cost-benefit analysis will allow one to compare the benefits of environmentally sustainable practices with the total cost of implementation.

13.8 Environmentally Sustainable Businesses also have a Competitive Edge

Environmentally sustainable businesses may also have a competitive edge when it comes to attracting customers and investors. Modern consumers are aware of social and environmental issues and keep themselves informed about which businesses are acting responsibly in the community. Investors are equally aware of these issues and there is a trend developing towards investing in environmentally sustainable companies.

Most importantly in considering the business case for environmental sustainability is the point that it doesn't negatively impact on a business' ability to generate a profit. In fact, in the long-term it is considered to improve profitability through the reduction of expenses and increased competitiveness.

These factors suggest that there is a business case for environmental sustainability. Each point can be capitalized on to generate returns and improve the bottom line in your organization. As more and more businesses implement environmentally sustainable processes into their operations, one put oneself at risk of being left behind if one don't actively get involved wherever one can.

13.9 Consumer Conscience and Public Image

Conscience is a person's ability to distinguish whether their actions are right or wrong based on basic moral values. When applied to the consumer, conscience refers to the way that people recognize which products is the most ethical choice when provided with a range of options. It is important that businesses understand the influence consumer conscience has on buying behaviour.

Consumer conscience is a significant contributor to the way that people choose which products they want to buy and from which business. It is influenced by a number of factors including popular opinion, the media, friends and family, government and personal values. At the moment, the protection of the environment is a particularly topical issue so it is something that consumers take into consideration. Consumers are now better educated about the impact that their actions and the actions of businesses can have on the environment.

13.10 Make them feel Good about Product

When people buy a product, or use a service that conforms to their personal moral values, they feel that they have done something good or have helped out as best they can. However, when consumers make a decision that goes against their principles, they may experience a sense of remorse or disappointment in themselves for making such a choice. Over time, these feelings shape the buying behaviour of customers, as people begin to actively seek out products that will make them feel good about themselves as well as contributing positively to the environment.

13.11 Capitalize on the Effects of Consumer Conscience

Businesses and marketers have recognized that there is an opportunity to capitalize on the effects of consumer conscience. If consumer conscience directly affects buying behaviour, then one should be able to tailor one's products and services to meet the expectations of your customers. Making changes to the way that one's business operates so that it is more environmentally sustainable is the first step.

Next, one should look at introducing a range of environmentally sustainable products so that consumers can make a choice about what they want to buy. When faced with the choice between an environmentally sustainable product and one that is not, many consumers will be swayed by their conscience into buying the sustainable option. One does not want one's product to be the one left on the shelf because it is damaging to the environment.

13.12 Environmental Sustainability in Business – A Significant Impact on Public Image

Businesses standpoint regarding environmental sustainability can have a significant impact on public image. Environmentally sustainable businesses are seen as market leaders, innovators and socially responsible. On the other hand, businesses that continue with unsustainable practices are viewed as outdated and as contributors to the destruction of the environment.

Establishing one's image as an environmentally sustainable business will help to build trust and respect from a broad range of consumers. One can use it to leverage one's business above the competition and gain a competitive edge, whilst at the same time making a positive contribution to the health and sustainability of the environment.

Without differentiation, all businesses that sell the same product or service would be in direct competition with one another. Business differentiation allows one to position one's products so that consumers can distinguish between those that are offered by one's business and those offered by the competition. The key is to differentiate one's business based on points that consumers value as important when making purchasing decisions.

13.13 An Emerging Trend is for Businesses Depends on Environmental Sustainability

A business can differentiate its products and services based on a range of different factors such as price, quality, usability, convenience and service. An emerging trend is for businesses to differentiate based on environmental sustainability. Businesses can choose to either offer a complimentary range of environmentally sustainable options alongside their existing products, or choose to make their entire operation more sustainable from the ground up.

13.14 Sustainability Differentiation Changes Consumer Expectations

The main driver behind environmental sustainability differentiation is changing consumer expectations. Consumers are become more aware and more interested in limiting their negative impact on the environment. Consumers are also openly expressing their disappointment in businesses that focus on profit making at the expense of a healthy environment. Businesses have identified that an opportunity exists to differentiate them based on their commitment to environmental sustainability.

However, before a business totally changes their proven business model to a new, sustainability focussed approach, they need to consider the potential risks and costs

involved. It can be difficult to successfully reposition a business or product in a competitive market, especially if they already have an established customer base.

13.15 Environmental Sustainability does not Damage or Remove any Part of One's Product

When making a move towards environmental sustainability, it is crucial that one does not damage or remove any part of one's product that one's customers value. For example, if one currently differentiate oneself on lowest price, one risk losing one's loyal customers if one suddenly changes to environmentally sustainable products that are high in price. Ideally, environmental sustainability should be an additional differentiating factor to one's existing product line and business model.

13.16 Marketing Strategy that Clearly Explains one's Commitment

If executed correctly, a marketing strategy that clearly explains one's commitment to a move towards environmental sustainability can be used to mitigate these risks. Many customers accept that there may be minor changes to products as business become more environmentally responsible. Additional information on packaging or an advertising campaign will help one to communicate your vision to one's customers and may also help one to attract new customers away from the competition.

Essentially, environmental sustainability differentiation should have two clear goals. Firstly, it should provide one with a competitive advantage or at least put one on equal ground with other businesses in the market. Secondly, it should help one to become more socially responsible and have a positive impact on the health of the environment.

13.17 Push towards Improving Environmental Sustainability – New Business opportunities

The current push towards improving environmental sustainability is creating many new business opportunities. Industries in the environmental sector that barely existed a few years ago are now emerging and growing at a rapid rate. There are numerous market opportunities for entrepreneurs, new enterprises and existing businesses to capitalize on in relation to environmental sustainability.

There is an increasing amount of investment being allocated to developing sustainable technology from both the private and government sector. This investment is helping to generate new innovations that offer environmentally sustainable solutions to long standing problems and is also driving the demand for employees in the sector.

13.18 Renewable Energy Sector

The renewable energy sector is one industry that has grown significantly as a result of increased interest in environmental sustainability. The demand for renewable energy is increasing as businesses and individuals begin to realize that it is a simple and practical way to reduce their overall negative impact on the environment. As governments move to legislate the proportion of energy that must come from renewable sources, this industry will continue to increase its share of the energy market.

Environmental sustainability consulting is another type of enterprise with excellent potential in the current social climate. Environmental sustainability consultants visit a business or home and analyse the total impact that it is having on the environment. They then offer advice and develop plans that you can implement to reduce your environmental impact and improve the efficient use of your resources. There is plenty of potential for consultants in this industry, as big businesses are prepared to spend on advice that offers them practical solutions to help reduce their environmental impact.

13.19 "Connecting People to Nature": World Environment Day, 5 June 2017

The Stockholm Conference that happened in first week of June in the year 1972 was the first global conference organized by the United Nations Organization, in which leaders from more than 196 countries across the world had participated. The major focus of this 10 days long conference was to deliberate upon the role of humans in causing deterioration of the nature and environment and how only the humans could contribute in mitigating such deterioration. This conference proved to be a major milestone in the history of the development of human concern about deteriorating conditions of the environment. Intense brainstorming sessions during this conference culminated in several treaties, decisions and protocols related to controlling and minimizing the pollution of the air, water and soil, protecting wildlife and biodiversity, conservation of the natural resources and so on and so forth. In this conference, it was also decided to observe 5th June every year as the World Environment Day to commemorate the conference and also to generate and spread awareness among the masses with regards to safeguarding the environment, natural resources, biodiversity, wildlife, forest ecosystems, etc., directly or indirectly through a series of activities, campaigns and other means. Since then, World Environment Day is celebrated every year with a central theme.

This year's theme is "Connecting People to the Nature" which is the most appropriate concept. Since the humans evolved on the planet Earth millions of years ago, they lived amidst the nature in the caves and the forests and respected the nature like every other living being. They had mingled with nature like every other animal living in the wild. So the human race was deeply connected and bonded with nature for several million years since their evolution. Eventually, the human civilization began to rise in 3000 BC. Although civilized, they were still connected with nature. The rise of Industrial Revolution in 18th through 19th century proved to be a major blow to this connectivity. Intimate bonding of the humans with the nature started breaking with ever-increasing human selfishness. The human activities became more and more development oriented for their own benefits and comfort taking a heavy toll on the health of the environment. It appeared as if humans and nature are drifting apart from each other like two straight lines running parallel to each other which would never meet each other! Increasing population, rapid urbanization, especially in the developing countries like India.

Source: Dr Sanjay Joshi, Department of Environmental Science, K. J. Somaiy College of Science and Commerce, Vidyavihar, Mumbai.
Director and one of the Founder Members, Enviro-Vigil, Thane.

13.20 Quotes from Different Personality

Nature has bestowed abundant wealth on human beings. Therefore, it is our fundamental duty to conserve nature with a sense of gratitude and social obligation. It is our collective responsibility to use natural resources efficiently, plant trees and protect wildlife and marine life.

Shri Devendra Fadnavis
Chief Minister, Maharashtra State

Global warming has caused anxiety all over the world. The need of the hour is to accept this challenge and make serious efforts to fulfil individual obligations. We are confident that proper environmental management will help resolve this crisis. It is absolutely necessary to support positive efforts of the public towards environment conservation and pollution control. Let us therefore be committed towards this cause.

Shri Ramdas Kadam
Minister for Environment, Maharashtra State

We must keep in mind that even small efforts can result in significant collective reform. We can do much more for preservation of natural resources by small acts such as avoiding misuse of water at domestic level and saving electricity. Determination and consistency are required to wholeheartedly pursue environmental gains.

Shri Pravin Pote Patil
Minister of State for Environment, Maharashtra State

Climate change and environmental degradation are serious challenges confronting the world today. However, the solution lies in awareness and positive action of society. Extensive awareness among school and college students combined with firm steps taken by diligent citizens will certainly yield results. It is still not late. Proper management of resources and active public engagement are positive steps towards preserving the environment.

Dr P. Anbalagan (IAS)
Member Secretary, MPCB

We need to conserve the environment as it is our social responsibility. Now this is time that we should think and act to minimize use of natural resources. I am sure with collective approach and active participation of people we could preserve healthy and green environment for the next generation.

Sumeet Mallik
Chief Secretary Govt. of Maharashtra

Each one of us must resolve to create a balanced environment through small individual initiatives and collective efforts. We must create environmental awareness among school children as they are the pillars for building a strong nation in future. Thus, it is necessary to emphasise protection and conservation of natural resources among the future generation. This is the fundamental rule for green reformation.

Satish Gavai (IAS)
Additional Chief Secretary, Environment Department/Chairman MPCB

13.21 Efforts taken by MPC Board, Mumbai to reduce Air Pollution in the State of Maharashtra during 2015–16

MPC Board enforced Air Pollution (Prevention & Control of Pollution) Act, 1981 in the State of Maharashtra. Also, implements the notifications published by the Ministry of Environment & Climate Change, New Delhi, Central Pollution Control Board and Environment Department, Govt. of Maharashtra. The mainly, Board deals the air pollution issues of the Metal and Mineral sectors. Board also prepared action plans, guidelines for the various sectors to reduce/control air pollution. Board has taken following efforts to reduce air pollution and noise level in the State of Maharashtra.

- Board issues Consents to the Metal and Mineral sector industries by incorporating conditions to install adequate air pollution control system to minimize the air pollution. Also, incorporate the condition to install advance specific air pollution control system such as Flue Gas Desulphurization to the Thermal Power Plants and Multiple Effective Evaporator for the metal bearing industries.
- Ministry of Environment Forest & Climate Change, Govt. of India, New Delhi was declared Comprehensive Environmental Pollution Index of the polluting clusters located in the State of Maharashtra. In that cluster Chandrapur, Tarapur, Aurangabad, Dombivali and Navi Mumbai region was declared as a polluted cluster/ cities. Due to high level of the pollution concentration upcoming, commissioning of the new projects were banned in the above mentioned regions.
- Board has taken efforts, studied all the pollution aspects/sources of the above regions and prepared the long-term and short-term action plans to reduce concentration of the air pollutants in that cities. Also, it was communicated to the concerned departments about implementation of action plans in cities to reduce pollution issues. Board has communicated the short-term and long-term action plans to the CPCB and MoEF, New Delhi along with implementation of the action plans with CEPI scores. After taking the regular follow-up and pursuing action taken report and its implementation to the MoEF & CC, New Delhi they have lifted moratorium by issuing memorandum from all above regions latest by Chandrapur region on 20th May, 2016.
- Also, Board is in the process to implement air pollution control action plan in the 18 cities of the State of the Maharashtra. To assess the ambient air quality in the cities of Mumbai, Pune, Solapur, Nagpur, Nashik, Chandrapur, Navi Mumbai and Aurangabad Board has installed Continuous Ambient Air Quality Monitoring Stations (CAAQMS). As well as Board has prepared action plan for 10 non attainment cities of the Maharashtra by issued work order to NEERI and IIT(B).
- Also Board has taken efforts to assess pollution potential of the clamped type traditional brick kilns, coal stack yards, tyre pyrolysis units and the stone quarries. After assessment of the pollution potential Board has formulated guidelines with the consultation of the Environment Department, Govt. of Maharashtra. These guidelines and notification were published on the Boards Website for public domain for effective implementation.

- Board has taken efforts to minimize noise level arising due to marriage hall/ gardens during the marriage functions and prepared the guidelines with the consultation of the Environment Department, Govt. of Maharashtra. This guideline was published on the Boards Website for public domain for the effective implementation.
- Board has established Air Care Centre at HQ, Sion, Mumbai for assessment of the air pollutants/air pollution level arising from the grossly polluting industries. This centre was inaugurated by the Hon'ble Chief Minister, Govt. of Maharashtra on 20/04/2016.
- Board has continuously taken efforts to reduce noise level in the State of Maharashtra by installation of the Noise Monitoring stations at various locations. Also, Board has issued work order to the NEERI, Nagpur for the monitoring and mapping of the noise levels in the 27 Municipal Corporation areas.
- Board has also installed Wind Augmentation and Air Purifying Unit (WAYU) at 5 traffic junction location in the Mumbai city namely at Sion, Bandra, Ghatkopar, Bhandup and Neat CST railway station to reduce air pollution arises from the transportation activity.
- Board has published Statistical Information of the Ambient Air Quality for the period of 2015–16 and 2016–17. Also, in the process to published this report in a Marathi language. We would like to inform you that all over India Maharashtra State have only published Air Quality report in respective years for public information.

Source: Hindustan Times marketing initiatives, World Environment Day, 5 June, 2017.

13.22 Resolution of the Protection of the Environment would prove to be Blessings for Prosperity

Since 1972, 5th June of every year is observed as World Environment Day. This year is no different and the concept being followed is "One Day with Nature". Changes in weather patterns were recorded for the first time on 5th June in Stockholm and ever since then the day has been observed as World Environment Day.

Almost five decades ago the entire world took note of the environment and if we compare the time we live in today to 1972, we would realize that even though we are on the path to development and prosperity we have completely ignored the deteriorating environment. Action back then would have limited the impact and we would not be facing an overwhelming challenge today.

The last 10 years have seen us over-populate mega-cities, and whatever efforts were made to limit the negative impact of harmful products, use of natural resources and purification of sewage water were clearly not enough. The reason being, a lack of effective planning and public involvement. This problem leads to larger consequences with every passing day.

After analyzing various issues of the environment, we have realized that the need to create public confidence on a large scale should be mandatory and should not solely be the responsibility of the Government. Changing lifestyles has seen us evolve in taste and preferences. Today the way we dress is different to our previous generation, we have accepted Pizza along with Vadapav, people prefer going to malls and departmental stores instead of visiting the good old grocery shops while the demand for air-conditioned public transport is only increasing. These demands are being met at a very high cost and this cost is proving fatal to the environment.

Consider this, the problem of plastic bags is present right from big metro-cities to small villages. To overcome a problem like this a small resolution to not use plastic bags by the public could bring a drastic change. But the lifestyle of today is to adopt shortcut methods such as the "use and throw" style and there in lays the problem. We see many senior citizens carrying cloth-bags and recyclable material which goes to show the older generation are aware of their responsibilities and are trying to make a difference. The younger generations are the ones who seem to be indifferent to environmental problems which are making situations worse. Small steps like switching off electricity when not in use, using cloth-bags for and recycling could create a noticeable change in the environment. But is it going to bring a change only if we do this? These type of questions are making the problem even more serious. For instance, one of the main triggers for the fatal 2005 Mumbai floods were plastic bags and human waste. After the floods had finally settled, public awareness grew and people became aware. However it did not take us too long to stop acting like moral citizens, forget our responsibilities and continue destroying the environment once again.

In order to resolve the current situation, we must take immediate action. Only once we have clean villages, clean towns and clean states only, will it be possible for our people to achieve our dream of a clean nation.

The Government has made laws for the protection of environment so that people can live their lives more comfortably, but these laws aren't necessarily implemented. In order for set systems to function, large scale public involvement, honesty and determination are essential. If everybody thinks in a manner that promotes social integrity and welfare these laws can be implemented with ease and the environment can begin to repair itself.

India is steadily moving towards becoming a superpower. But unless we carefully manage land, natural resources and water resources responsibly, our dream to become a superpower will remain merely a dream.

Therefore it is our duty as citizens of India to act responsibly and keep our country clean, not just our homes, to save water whenever possible and to adopt the faith that worshiping nature is like worshiping God. We should not forget our ancient culture, traditions, history, national integrity and freedom fighters, they make us who and what we are today. A nation which thinks about their future in the present by remembering their history is one which is able to sow the seeds of a bright future for younger generations. As citizens of not only this prosperous country, but as citizens of the world we must try to protect our environment. We must try to make a change every day and not forcefully only on days of significant importance. The Earth is our home, it gave us love and life and now the time has come to give back Mother Nature what it has given us.

Source: Hindustan Times marketing initiatives, world Environment Day, 5 June, 2017.

13.23 Conclusion

The key to capitalizing on business opportunities in the environmental sustainability industry is to create custom solutions tailored to client's individual situation. Different businesses will have different visions of what they want to achieve and how much they are prepared to change and spend to become more environmentally sustainable. By offering custom solutions, one can differentiate oneself from the generic options available in the market. Hence let us conserve the green earth and preserved the enriched environment.

Government incentives offered to businesses that improve their environmental sustainability also offer a unique business opportunity. There are grants, funding and tax incentives available for start-up enterprises in the environmental sustainability sector as well as for businesses that actively work at reducing their environmental impact. One can find further information about these incentives from Australian government's department of the Environment, Water, Heritage and the Arts website.

Section 4

Environmental Sustainability and Economic / Social Progress

CHAPTER 14

Strategies for Environmental Protection

14.0 Overview

There has been an increasing awareness in recent years that protection of the environment is necessary for sustaining the economic and social progress of country. This awareness was reflected at the Earth Summit in Rio de Janeiro in June 1992, where more than a hundred heads of government adopted a global action plan called Agenda 21 aimed at integrating environmental imperatives with developmental aspirations and reiterated through the UN. General Assembly Special Session on Environment held in June 1997. It is now accepted that, in terms of natural resources, a country's demand for its sustenance should not exceed its carrying capacity. Over the last few decades, India has evolved legislations, policies and programmes for environmental protection and conservation of natural resources.

The Indian Government's policy has been expressed in the form of statement on forestry, on the abatement of pollution, the national conservation strategy and the policy statement on environment and development. The spirit of Agenda 21 principles has already been incorporated in these policies. For instance, with regard to the social and economic dimensions of Agenda 21, India has become a signatory to the protocol for phasing out ozone depleting substances, the Basel Convention on trans-boundary hazardous substances, the Convention on biological diversity and other international treaties. Similarly, poverty alleviation programmes have been launched wherein family planning and welfare is a major focus. Environmental concerns are being integrated with development in decision making through mandatory clearance of projects based on environmental impact assessment. Compliance with the conditions stipulated is being ensured by monitoring the progress of implementation of Environmental Management Plans.

Conservation and management of resources for development are sought to be achieved through a combination of regulatory and market-based economic instruments. The role of major groups including the NGOs, farmers and other communities is being strengthened by directly involving them in the process of identification, formulation and implementation of environmental programmes. The important role of capacity building, legal instruments and mass media for promoting public awareness is fully recognized.

14.1 Policy on Environment

The Indian Government's policy towards environment is guided by the principles of Agenda 21. The Government of India has issued policy statements on:
(i)   Forestry,
(ii) Abatement of Pollution,
(iii) National Conservation Strategy,
(iv) Environment and Development,
(v)   National and Development India is already a signatory to the Convention on Biological Diversity, Montreal protocol and Basel Convention.

Global environmental issues, such as ozone depletion, climate change due to accumulation of Greenhouse Gases (GHGs), biodiversity loss, etc., are largely due to the rapid industrialization of the developed nations. India is an insignificant contributor to the GHG emissions as can be seen from the Table 14.1.

Table 14.1:   Carbon Emission Levels in Selected Countries*
(million tonnes)

Country	Estimated Share of World Population 2014	Estimated Share of Gross World Product 2014	Estimated Share of World Carbon Emissions 2013	Emission per capita
United States	5	26	23	5.3
Russia	3	2	7	2.9
Japan	2	17	5	2.4
Germany	1	8	4	2.9
China	21	2	13	0.7
India	17	1	4	0.3
Indonesia	4	1	1	0.3
Brazil	3	2	1	0.4
Total	56	59	58	0.9

*Compilation from several international published sources.

The main environmental problems in India relate to air and water pollution, degradation of common property resources, threat to biological diversity, solid waste disposal and sanitation. Increasing deforestation, industrialization, urbanization, transportation and input-intensive agriculture are some of the other major causes of environmental problems being faced by the country. Poverty presents special problems for a heavily populated country with limited resources.

14.2 Status of India's Environment

14.2.1 Air Quality

The urban areas represent complex environmental problems. The living conditions of millions of urban poor are such that they pose a threat to their health and have potentially catastrophic social consequences. For the urban poor, the living conditions are the worst. If these problems are not addressed to in an adequate and timely manner, serious environmental and associated health consequences will follow. Burgeoning urban population beyond the carrying capacity of the different components of urban eco-systems, coupled with indifferent urban governance, are the root causes for urban environmental problems. Air pollution can cause chronic and acute respiratory diseases, ventilatory malfunction, heart diseases, cancer of the lungs and even death.

The blood lead levels of persons in Ahmedabad, Bombay (now known as Mumbai) and Kolkata have been reported to be higher than the corresponding levels of persons in lead-free gasoline areas. The details of ambient air quality status in ten large cities/towns are at Annexure 14.1. In most of the cities, while the SPM levels of SO_2 and NOx are significantly higher than the CPCB standards. The rural population uses substantial quantities of non-commercial fuel i.e. crop residues, animal dung or wood. Although their share in total fuel consumption is decreasing, these still provide 80 per cent of rural energy for cooking. Several adverse health effects are suspected to arise due to indoor pollution especially where conventional "sigri" has not been replaced with smokeless *chulha*. Respiratory infection in children, chronic lung disease, lung cancer in adults and adverse pregnancy outcomes, such as low birth weight and still birth of the child, for women exposed during pregnacy, are some of the diseases associated with indoor pollution.

14.2.2 Water Resources and Quality

India is rich in water resources, being endowed with a network of rivers and vast alluvial basins to hold groundwater. Besides, India is blessed with snow cover in the Himalayan range which can meet a variety of water requirements of the country. However, with the rapid increase in the population of the country and the need to meet the increasing demands of irrigation, human and industrial consumption, the available water resources in many parts of the country are getting depleted and the water quality has deteriorated. In India, water pollution comes from three main sources: domestic sewage, industrial effluents and run-off from agriculture.

14.2.3 Public Health in both Rural and Urban India

The most significant environmental problem and threat to public health in both rural and urban India is inadequate access to clean drinking water and sanitation facilities. Almost all the surface water sources are contaminated and unfit for human consumption.

The diseases commonly caused by contaminated water are diarrhoea, trachoma, intestinal worms, hepatitis, etc. Many of the rivers and lakes are getting contaminated from industrial effluents and agricultural run-off, with toxic chemicals and heavy metals which are hard to remove from drinking water with standard purification facilities. Even fish and shellfish in such water get contaminated and their consumption may cause diseases.

14.2.4 Protection of Environment to Union and State Governments – The Constitution of India

The constitution of India has assigned responsibility of protecting the environment to the Union and State Governments. Environmental protection laws have been enacted under the Environment (Prevention and Control of Pollution) Act, 1986; the Air (Prevention and Control of Pollution) Act, 1981 and the Water (Prevention and Control of Pollution) Act, 1974 has maintained the ambient air and water quality standards, to demand information regarding effluent emissions, to shut down polluting activities and to prevent discharges of effluent and sewage. Although these regulations have given the CPCB and its State-level counterpart's broad powers to control the problem of air and water pollution, the enforcement has been weak.

14.3 Solid Wastes and Hazardous Chemicals

There has been a significant increase in the generation of domestic, urban and industrial wastes in the last few decades. This is largely the result of rapid population growth and industrialization. The per capita solid waste generated is estimated at 0.20 tonnes in Mumbai, 0.44 tonnes in Delhi and 0.29 tonnes in Chennai. Although a major part of the waste generated is non-hazardous, substantial quantities of hazardous waste is also generated.

14.3.1 Chemical Industries Release Huge Quantities of Wastes into the Environment

The growth of chemical industries has resulted in the extensive use of chemicals, which release huge quantities of wastes into the environment in the form of solid, liquids and gases. A substantial amount of these wastes are potentially hazardous to the environment. The leaching of hazardous wastes at the dumping sites is a common feature. This result in the contamination of surface and groundwater supply and is a potential risk to human health. Effective control of hazardous wastes is of paramount importance for the maintenance of heath, environmental protection and natural resource management.

14.3.2 Hazardous Wastes (Management and Handling)

Hospital wastes being generated by mushroom growth of nursing homes pose a special risk and has the potential to take epidemic form. In view of the proliferation of the chemical industry and the significant increase in the hazardous waste generation, the Government of India framed the Hazardous Wastes (Management and Handling) Rules time to time. Under these Rules, it is mandatory for the hazardous waste generators to provide information on the quantity and type of hazardous wastes produced.

14.4 Land Degradation and Soil Loss

Soil erosion is the most serious cause of land degradation. Estimates show that around 130 million hectares of land (45% of total geographical area) is affected by serious soil erosion through ravine and gully, cultivated wastelands, water logging and shifting cultivation. It is also estimated that India loses about 5310 million tonnes of soil annually. The accumulation of salts and alkalinity affect the productivity of agricultural lands in arid and semi-arid regions, which are under irrigation. The magnitude of water logging in irrigated command has recently been estimated at 2.46 million hectares. Besides, 3.4 million hectares suffer from surface water stagnation. Injudicious use of canal water causes water logging and a rise in the water table, which, if left uncorrected eventually leads to salinization. Although irrigation and drainage should go hand in hand, the drainage aspect has not been given due attention in both major and minor irrigation projects in the country. There has been water logging associated with many of the larger reservoirs since their inception.

Fertilizers and pesticides are important inputs for increasing agricultural production. Their use has increased significantly from the mid-60s. Over and unbalanced use of these chemicals is fraught with danger. However, fertilizers and pesticides use are concentrated in certain areas and crops. Suitable agronomic practices will be helpful in this regard.

Table 14.2 and Table 14.3 show that our consumption of pesticides and fertilizers is much below that of the neighbouring countries.

Table 14.2:   Trends in the Consumption of Chemical Fertilizers in Selected Asian Countries

Country / Year	Use of Chemical Fertilizers per Hectare of Arable land (kg/ ha)			
	1970–71	1991–92	2001–02	2013–14
Bangladesh	15.7	109.8	89.9	67.8
Bhutan	0.8	0.8	0.7	0.5
India	13.7	75.2	65.5	43.2
Nepal	2.7	27.2	20.5	11.2
Pakistan	14.6	88.9	58.3	36.9
Sri Lanka	55.5	93.7	68.5	45.3
Philippines	28.7	54.8	40.2	25.4
China	41.0	304.3	145.5	85.5
Rep. Korea	245.0	451.7	210.2	121.5
Japan	354.7	387.3	155.4	103.5

Source: Compiled from Annual Reports of various countries.

Table 14.3:   Average Level of Consumption of Pesticides in Different Countries

S. No	Country	Level of Consumption (kg/ha)
1.	Argentina	0.295
2.	India	0.450
3.	Turkey	0.298
4.	Indonesia	0.575
5.	U.S.A	0.579
6.	Thailand	1.367
7.	Mexico	1.375
8.	Republic of Korea	6.559
9.	Japan	9.180
10.	Hungry	12.573
11.	Italy	13.355

Source: Compiled from Annual Reports of various countries.

14.5 Forests, Wildlife and Biodiversity

Forests are important for maintaining ecological balance and preserving the life supporting system of the earth. They are essential for food production, health and other aspects of human survival and sustainable development. Indian forests constitute 2 per cent of the world's forest area but are forced to support 12 per cent of the world's human population and 14 per cent of world's livestock population. This is sufficient to indicate the tremendous biotic pressure they face.

Forests in India have been shrinking for several decades owing to the pressure of population on land for competing uses such as agriculture, irrigation and power projects, industry, roads, etc. In India, forests account for about 19.27 per cent of the total land area. On the other hand, in advanced countries, the area under forests is often about a third of the total land area. There is a need to have massive reforestation programmes, control over hacking and grazing and grazing and provision of cheap fuel alternative technologies.

14.6 The National Forest Policy

The National Forest Policy 1988 stipulates that a minimum of one-third of the total land area of the country should be brought under forest or tree cover. It is envisaged that this will be achieved by involving local stakeholders like the farmers, the tribals, the women, the NGOs and the Panchayat Raj Institutions (PRIs). Another concern relating to the state of forest resources is that of biodiversity and extinction of species. India has a rich heritage of species and genetic strains of flora and fauna. Out of the total eighteen biodiversity hot-spots in the world, India has two, one is the north-east Himalayas and the other is the Western Ghats. At present, India is home to several animal species that are threatened, including over 77 mammal, 22 reptiles and 55 birds and one amphibian species. For in-situ conservation of biological diversity, India has developed a network of protected areas, including national parks, sanctuaries and biosphere reserves. This network, which is being progressively expanded, now covers about 4 per cent of the total land area of the country. Because of the amendments in 1991 to the Wildlife (Protection) Act, hunting of all species of wildlife for commerce or for pleasure has been banned.

14.7 Highlights in the Achievements of Environment Sector

Environmental protection covers all those activities which relate to the formulation of policies and programme for prevention and mitigation of pollution through the regulatory framework. Besides, activities which are initiated for the conservation of ecology are also included. An amount of ₹525.000 crore was allocated for the environment sector in the Twelfth Plan. This planning commission subsequently is replaced by National Institute for Transforming India (NITI) Aayog by the National Democratic Alliance (NDA) Government in 2014. The Aayog has started working in this line[*]. The major highlights of the achievements in the environment sector during the Eighth Plan are as below.

14.8 Abatement of Pollution

14.8.1 Central Pollution Control Board

The main functions of Pollution Control Board are to act as regulatory agency for the prevention and control of water and air pollution by invoking, wherever necessary, the Water (Prevention and Control of Pollution) Act, 1974 and its subsequent amendment, the Air (Prevention and Control of Pollution) Act, 1981 with amendments from time to time and the Environment Protection Act, 1986 and its various amendments to supervise the work of the State Pollution control Boards. The major activities of the Central Pollution Control Board during the Eighth Plan included the development and expansion of laboratory facilities, management and operation of the national air and water quality network, controlling pollution at sources, river basin studies, evaluation and implementation of national standards, hazardous waste generating industries in different States, preparation of Zoning Atlas for siting industries in various districts of the country, development of criteria for eco-labelling of consumer products, remedial measures for vehicular pollution especially for vehicles in use in metro cities, noise pollution survey, training of personnel engaged in preventing and controlling pollution and organizing nationwide awareness programmes for prevention and control of pollution.

---

[*] Separate chapter i.e. Chapter 17 has been added to know the aims, objectives and achievement of NITI Aayog.

14.8.2 Environment Statement (as part of Environmental Audit)

Submission of an environmental statement by the polluting units to the concerned State Pollution Control Boards has been made mandatory through a gazette notification issued under the environment (Protection) Act – 1986. The environmental statement enables the units to take a comprehensive look at their industrial operations and facilitates an understanding of material flows and focussing on those areas where waste reduction, and consequently saving in input costs, is possible.

14.8.3 Adoption of Clean Technologies in Small Scale Industries

This scheme seeks
  (a) to promote the development and adoption of clean technology, including waste reuse and recycling and
  (b) to link research and development with dissemination of the R&D outcome and adoption of clean technologies to prevent pollution in small scale industries. Activities relating to demonstration of already proven cleaner technologies/techniques, small scale industries and waste minimization and demonstrating studies in selected sectors were undertaken during the Eighth Plan.

The concept of Waste Minimization Circles is as follows:
  (i) Promotion of the concept of waste minimization through awareness and training programmes.
  (ii) Institutionalization of waste minimization circles among the clusters of small scale industries of the same category.
  (iii) Sectoral studies on waste minimization and demonstration in selected.
  (iv) Demonstration improvement in environmental and in turn economic performance in a cluster of small units.
  (v) Promotion of increased general environment awareness among small scale units by providing training programmes on various environmental issues.
  (vi) Preparation of sector-specific manuals on waste minimization.
  (vii) Preparation of training packages on waste minimization and organization of training programmes for trainers and trainees.

14.8.4 Environmental Statistics and Mapping

Under this scheme, activities relating to collection, collation and analysis of environmental data and its depiction on an atlas were carried out. Activities relating to the production of computerized maps and preparation of Zoning Atlas for siting industries in selected districts were also taken up during the Eighth Plan.

14.8.5 World Bank Assisted Industrial Pollution Control Project (Phase-I)

This project has the following two broad components:
  • Investment Component which provides for loan assistance to large and medium scale industries for installing pollution control equipments; establishment of common effluent treatment plants for clusters of small scale units; and establishment of

demonstration projects for introducing energy and resource conservation measures in the small and medium scale sectors.

- Institutional Development Component designed to strengthen the monitoring and enforcement abilities of the Pollution Control Boards of four industrialized States of Gujarat, Maharashtra, Tamil Nadu and Uttar Pradesh. These include activities like acquisition of analytical and monitoring equipments, provision of laboratory facilities and training.

The medium and large-scale industries have utilized the loan amount disbursed to them by the IDBI and ICICI. As many as 35 common effluent treatment plants have been extended financial assistance. About 100 training programmes have been conducted for the personnel of the Central and State Pollution Control Boards. A dozen demonstrations projects have been approved for different technologies to be developed in various industries. The equipment for the identified State Pollution Control Boards under Phase-I have been partly procured.

14.8.6 World Bank Assisted Industrial Pollution Prevention Project (Phase-II)

This project has the following objectives:
- (i) to strengthen the capabilities in the States of Rajasthan, Madhya Pradesh, Karnataka and Andhra Pradesh,
- (ii) to facilitate priority investments to prevent pollution from industrial sources by encouraging the use of clean technologies, waste minimization and resource recovery,
- (iii) to provide technical assistance for the adoption of modern tools of information and management, organization of clean technology institutional network and an extension service on environmentally sound practices for small scale industries. The project has three components, namely institutional, investment and technical.

14.8.7 Development of Standards

Development of standards is a continuous process and they are notified as and when they are finalized for specific categories of industries.

14.8.8 Industrial Pollution Control

The activities under Industrial Pollution Control programme includes:
- (a) Taking priority action to control industrial pollution for which 17 categories of heavily polluting industries in the country have been identified and a time-bound programme has been given to the industries to install necessary pollution control facilities and to comply with the prescribed standards. The follow-up action on compliance is being monitored.
- (b) Monitoring of action points relating to restoration of environmental quality in critically pollution areas.
- (c) Preparation of Zoning Atlas for siting industries.
- (d) Implementation of pollution control measures in Agra-Mathura region.

14.8.9 Pollution Monitoring and Review

The activities under this programme include:
(i) monitoring of coastal water, river water and groundwater quality,
(ii) assessment of coastal pollution to plan for its prevention,
(iii) ambient air quality monitoring and
(iv) industrial inventory for large, medium and small scale industries.

14.9 Economic Instruments

In an effort to integrate economic and environmental planning, a variety of incentives to adopt efficiency enhancing and waste minimization practices are being promoted. This includes enhancing the cess rates on water consumption, duty concession, accelerated depreciation on pollution abatement equipment, etc. To facilitate a wider introduction of such instruments, a study has been sponsored by the Ministry of Environment and Forests to analyse market-based instruments such as taxes/charges for industrial pollution abatement.

14.10 Economic and Environmental Planning

Government policies, in addition to regulatory mechanisms, incorporate market-based economic and environmental planning. For example:
(a) Enhancement of cess rates on water consumption.
(b) Duty concessions on import of certain pollution control equipments.
(c) Accelerated depreciation on pollution abatement equipment.

14.10.1 Environmental Impact Assessment (EIA)

The purpose of Environmental Impact Assessment is to appraise developmental projects to ensure that development takes place in harmony with environmental concerns. It also enables the project authorities to integrate environmental concerns in the project portfolio. In a way it is a preventive measure. Other related activities carried out during the Eighth Plan included: carrying capacity studies (Doon Valley, National Capital Region), studies on improving the methodology and techniques of environmental impact assessment of development projects, training programmes, promotion of cleaner production programmes, including life cycle studies.

14.10.2 Conservation and Survey

14.10.2.1 Botanical Survey of India (BSI)

About 65 per cent of the total area of the country has been surveyed and three million herbarium specimens are in possession. More than 106 new species were discovered by BSI. Surveys in special/fragile ecosystems like cold deserts, hot deserts, Alpine Himalayas, wetlands, mangroves and coastal areas have been undertaken. The BSI also undertook special projects such as a study on conservation and survey of rare and endangered species, all-India coordinated project on ethnobiology, floristic study of biosphere reserve areas, EIA in developmental project areas, geo-botanical studies in Singhbhum and Khetri copper belts, etc.

14.10.2.2 Zoological Survey of India (ZSI)

About 65per cent of the total area of the country has been surveyed. About 759 new species were recorded. The main activities of ZSI during the Eighth Plan were: exploration and survey of faunal resources, taxonomic and ecological studies, maintenance and development of national zoological collections, status survey of endangered species, environmental impact assessment studies, publication of Fauna of India.

14.10.2.3 National Museum of Natural History (NMNH)

The NMNH, New Delhi is an institution devoted to environmental education. The highlights of the Eighth Plan performance of NMNH relate to "LEARN" (Lessons on Environmental Awareness and Resources at NMNH) for the students of classes VI to XII of Delhi schools; "Environment Essay Competition" (in Braille) and "Feel, Smell and Tell" for visually handicapped; and "Know About Dinosaurs" for teenagers.

14.10.2.4 Biodiversity Conservation

The scheme on Biodiversity Conservation was designed to ensure proper coordination among various agencies concerned with the issues relating to conservation of biological diversity and to review, monitor and evolve adequate policy instruments for the same. The Convention on Biological Diversity (CBD) was signed by 168 countries, including India, during the Rio meetings. India has since ratified the Convention.

14.11 Research and Development

This is a continuing scheme for the promotion of research in the multidisciplinary aspects of environmental protection, conservation and development together with the creation of facilities and development of technical capabilities. To achieve these objectives, the research projects in the thrust areas are supported with grants-in-aid. Under the scheme, programmes such as Man and the Biosphere Programme, Environmental Research Programme, Action-Oriented Research Programme on Eastern and Western Ghats and research projects in climate change are included. Over 190 research projects in multi-disciplinary aspects were initiated. Of these, about 60 have been completed and results disseminated to the potential/interested user agencies.

A Status Report on the All India Coordination Project on Ethonobiology, documenting information of the country-wide survey concerning traditional knowledge system, use of biological resources by the tribal population and their interdependence, has also been published. A report on the All India Coordinated Project-III on conservation of endangered plant species has been published. A coordinated research project on Aerobio-pollution and Human Health was launched to collect information through survey, concerning air-borne diseases involving 26 centres throughout the country. Attention has also been paid to initiate studies on improving our understanding of the subject of climate change and on preparation of inventories of the greenhouse gases, which would be useful in projecting the scientific inputs for various discussions at the international level, as also to meet the requirements of the general commitment emerging out of the Framework Convention on Climate Change. A new scheme on environmental information dissemination was launched to ensure public participation in the programmes of environment awareness generation, control of pollution and conservation of natural resources. The scheme, named "Paryavarna Vahini" is being implemented. About 184 districts in various states of the country were selected to set up "Paryavaran Vahinis".

14.12 Environmental Education, Training and Information

In order to encourage participation of school children in various activities related to ecological conservation and preservation of the environment a scheme namely, Eco-clubs involving school children has been launched. The objective of the Eco-clubs is not limited only to imparting environment education to school children but also includes mobilizing them to participate in various environmental preservation efforts in their locality. More than 5,000 such Eco-clubs have been set up in various schools of the country.

14.12.1 An Environmental Information Systems (ENVIS)

It was set up by the Ministry of Environment and Forests to provide information on various subjects related to environment to decision-makers, researchers, academicians, policy planners, environmentalists, engineers and the general public. It is a decentralized system with a network of distributed subject-oriented centres, ensuring integration of national efforts in environmental information, collection, collation, storage, retrieval and dissemination to all user groups.

14.12.2 Chain of Distribution Centres

A chain of 22 such distribution centres, known as ENVIS centres, were set up on various priority areas of environment under the scheme, by the end of Eighth Plan. Five Centres of Excellence in the field of environmental education, ecological research, mining, environment and ornithology have been set up. These centres provide various resource materials, training, and research facilities, etc., to all concerned. As per the recommendations of the Standing Committee on Bio-resources and Environment, 37 priority areas were identified for undertaking research projects involving remote sensing technologies under a scheme of National Natural Resource Management System. Projects covering more than 18 areas were considered by the Bio-resource Committee and sanctioned during the Eighth Plan. Financial assistance has also been provided for the organization of seminars/symposia/workshops on environment related topics of scientific interest and to provide a common platform to all professionals for sharing the updated knowledge on environmental-related areas.

14.13 Policy and Law

The Government of India has enunciated its policy, in the form of policy statements, on Abatement of Pollution, on Forestry and National Conservation Strategy and on Conservation and Development. In addition, there are laws for protection of environment. These include Wildlife (Protection) Act, 1972; Forest (Conservation) Act, 1980; Water (Prevention and Control of Pollution) Act, 1974; Air (Prevention and Control of Pollution) Act, 1981; Environment (Protection) Act, 1986; Public Liability (Insurance) Act, 1991; and National Environment Tribunal Act, 1995. The Environment (Protection) Act, 1986 sets out the parameters under which the Ministry of Environment and Forests operates to the policy statement is the recognition of the principle that effective management and control of natural resources requires the support and participation of the people. Considerable attention was given to make the protection of environment by Government of India and various state governments.

14.13.1 Effective Implementation of the Environment Protection Act

A number of Central and State executive authorities have been delegated powers for effective implementation of the Environment Protection Act. The Air (Prevention and Control of Pollution) Act, 1981 and the Water (Prevention and Control of Pollution) Act, 1974 have been amended to bring certain provisions of the Acts at par with those of the Environment (Protection) Act, 1986. The National Environment Tribunal Act, 1995, provides for strict liability for damages arising out of any accident occurring while handling any hazardous substance and for the establishment of a National Environment Tribunal for effective and expeditious disposal of cases arising from such accidents with a view to granting relief and compensation for damages to persons, property and the environment and for matters connected therewith or incidental thereto. All the Acts were subsequently amended from time to time to make fine tuning.

14.14 International Cooperation

The Government of India participated in the conventions on implementing the Rio Agreements and the Agenda-21, Montreal Protocol, Commission on Sustainable Development, Global Environment Facility. The Indo-Canada Environment Facility is a commodity grant from the Canadian Government.

The grant is in the form of Murate of Potash, which is sold in the Indian market and the proceeds passed on, through the Ministry of Environment and Forests, to a registered society to undertake projects on environmental protection and conservation. India became a party to the United Nations Framework Convention on Climate Change (UNFCCC) Convention on Biological-diversity (CBD), Basel Convention on the Control of Trans-boundary Movement of Hazardous Wastes and their Disposal and Montreal Protocol on controlling the substances that deplete the Ozone layer.

14.15 National River Conservation Programme

14.15.1 Ganga Action Plan – Phase I

The Ganga Action Plan (GAP) Phase-I was launched by the Government of India in June 1986 as a 100 per cent Centrally Sponsored Scheme with the objective of improving the river water quality. It was envisaged that industrial pollution would be tackled through the enforcement of existing regulation by municipal authorities under which effluent treatment plants would be set up by industry. Under GAP-I, interception, diversion and treatment of sewage works; electric crematoria; low-cost sanitation and river front facilities were set up. The major reasons identified for the slippages have been problems in land acquisition, related litigation and contractual issues. As many as 683 mld of sewage treatment facilities have been extended against the target of 873/882 (revised). The scheme has been subjected to technical evaluation by four universities located on the banks of the Ganga. An ex-post evaluation in the "Benefit Cost Analysis" framework is in progress and the final report is awaited. As Biochemical Oxygen Demand (BOD) is a measure of the amount of organic pollution in water, it serves as a useful parameter for assessing water quality. Maximum success in reducing pollution in the river Ganga has been achieved in Allahabad followed by Varanasi and Kanpur.

14.15.2. Ganga Action Plan – Phase II

The GAP Phase II was launched during the Eighth Plan. Works on the major polluted tributaries of Ganga, namely, Yamuna, Gomati and Damodar, were taken up with the objective of improving the river water quality, as per the designated best use criteria. Works in 29 Class-1 towns along the Ganga, which could not be included in the first phase, were taken up in Phase II together with works in other smaller towns along the Ganga. The Scheme was launched as a centrally sponsored scheme with equal sharing by the Central and State Governments with the operation and maintenance expenses being fully borne by the States. The Yamuna Action Plan and the Gomati Action Plan components were approved. Work was taken up on Damodar River also. Summer average values for water quality (Dissolved Oxygen, BOD) on main stem of the river Ganga under GAP were taken into consideration.

14.15.3. National River Conservation Plan (NRCP)

The NRCP, envisaged the coverage of 18 grossly polluted stretches of rivers in 10 states. As many as 46 towns are to be covered of which 17 are in the southern, 11 in the western, 7 in the eastern and 11 in the central part of India. The NRCP was launched as a centrally sponsored scheme. The total cost of the scheme has been placed at crore and the time-frame for its completion is 10 years. The towns to be included in NRCP are given in Table 14.4. The GAP-II and the National River Conservation Plan were approved as centrally sponsored schemes with a sharing cost of 50:50. Through a government resolution, GAP-II merged with NRCP to cover a total of 141 towns on 22 rivers stretches in 14 states. The operations and maintenance (O&M) under the NRCP has not been found to be satisfactory. Lack of interest by the local bodies in the maintenance of sewage system, etc., and problems of uninterrupted power supply to sewage treatment plants, pumping stations, electric crematoria, etc., have been found to be the main causes. Unless O&M facilities created are improved, optimum benefits of the project cannot be achieved.

Table 14.4: Towns to be Covered under NRCP

S. No	Town	S. No	Town	S. No	Town
ANDHRA PRADESH		MADHYA PRADESH		34.	Chandbali
1.	Mancherial	17.	Indore	35.	Dharamshala
2.	Bhadrachalam	18.	Ujjain	PUNJAB	
3.	Rajamundry	19.	Burhanpur	36.	Ludhiana
4.	Ramagundam	20.	Mandideep	37.	Jalandhar (earlier known as Jullundur)
BIHAR		21.	Bhopal	38.	Phagwara
5.	Ranchi	22.	Vidisha	39.	Phillaur
6.	Jamshedpur	23.	Jabalpur	RAJASTHAN	
7.	Ghatshila	24.	Seoni	40.	Kota
GUJARAT		25.	Chapara	41.	Keshoraipatan
8.	Ahmedabad	26.	Keolari	TAMIL NADU	
KARNATAKA		27.	Nagda	42.	Kumarapalayam
9.	Shimoga	MAHARASHTRA		43.	Bhavani
10.	Harihara	28.	Karad	44.	Erode
11.	Bhadravati	29.	Sangli	45.	Trichy
12.	Davanagere	30.	Nasik	46.	Pallipalayam
13.	K.R.Nagar	31.	Nanded		
14.	Kollegal	ORISSA			
15.	Nanjangud	32.	Cuttack		
16.	Srirangapatna	33.	Talcher		

Source: Plan Documents, GOI.

14.16 Forests

The assessment of the forest cover of India based on visual and digital interpretation of the satellite data on a scale of 1: 250,000 the forest cover of the country is only 19.27 per cent of the total geographic area. Goa, Gujarat, Haryana, Himachal Pradesh, Jammu and Kashmir, Karnataka, Maharashtra, Mizoram, Punjab, Rajasthan, Sikkim, Tamil Nadu, Tripura, Uttar Pradesh and West Bengal have shown an improvement in the forest cover, whereas Andhra Pradesh, Arunachal Pradesh, Assam, Bihar, Kerala, Madhya Pradesh, Manipur, Meghalaya, Nagaland, Orissa, and the Union Territories of Andaman & Nicobar Islands have shown a further deterioration of forest cover. In Delhi, Chandigarh, Dadra and Nagar Haveli, and Daman and Diu, there was no change in forest cover during the period of last two assessments. On an aggregate basis, there has been a reduction in the forest cover to the extent of 5,482 sq. km between the two assessments of 2014 and 2015. Of the greatest concern is the picture in the North-Eastern States where a reduction to the extent of 783 sq. km. in the forest cover was reflected in the assessment.

The recent assessment shows a somewhat better situation, as the loss of forest cover in this region has come down to 316 sq. km. Mizoram and Tripura have, in fact, shown gain in forest cover. Under forest protection and regeneration, the scheme "Association of Scheduled Tribes and Rural Poor in Regeneration of Degraded Forests" was taken up on pilot basis with 37 projects in nine states namely Andhra Pradesh, Bihar, Gujarat, Madhya Pradesh, Maharashtra, Rajasthan, Orissa, West Bengal and Karnataka. A centrally sponsored scheme, "Modern Forest Fire Control Methods in India", was continued. The scheme was launched with UNDP assistance as a pilot project in Uttar Pradesh and Maharashtra mainly to protect forest from fire. The project was implemented in 13 states during Eighth protect forest from fire. The project was implemented in 13 states.

14.17 Afforestation and Eco-Development on Degraded Forests

The National Afforestation and Eco-Development Board was created at the time of bifurcation of the erstwhile National Wasteland Development Board, then under the Ministry of Environment and Forests. Areas adjoining forests and fragile eco-systems were brought under the National Afforestation and Eco-Development Board (NAEB) while other wastelands were covered under the newly-created National Wasteland Development Board in the Department of Waste and Development in the Ministry of Rural Areas and Employment. A considerable amount was alloted to NAEB for reclaiming degraded forest areas and adjoining forests through the following schemes:

14.17.1 Integrated Afforestation and Eco-Development Projects Scheme (IAEPS)

This is intended to promote afforestation and development of degraded forests by adopting an integrated watershed-based approach. This 100 per cent centrally sponsored scheme envisages micro-plan preparation by a multi-disciplinary team in consultation with the local people. Under this scheme an area of about 2,89,917 ha. was covered.

14.17.2 Fuelwood and Fodder Project Scheme

This is meant to augment the production of fuelwood and fodder in 229 identified fuelwood deficient districts of the country to meet the needs of the communities. The cost of raising the plantations of fuelwood and fodder is shared equally between the Central and the State Governments. Under this scheme an area of about 3,87,216 ha. was covered.

14.17.3 Non-Timber Forest Produce Scheme

The scheme provided for financial assistance to State Governments for increasing the production of Non-Timber Forest Produce (NTFP), including medicinal plants by raising plantations. This 100 per cent centrally sponsored scheme has a focus on creation of NTFP plantation assets in tribal areas. During the Eighth Plan period an area of about 1, 06,170 ha. was covered with a total expenditure of ₹56.47 crore under this scheme.

14.17.4 Grants-in-Aid Scheme

Promotion of people's participation in afforestation activities is a mandate of the NAEB. Under this scheme, non-governmental organizations (NGOs) are assisted financially for taking up afforestation and tree planting in public and private wastelands adjoining forest areas and building upon people's movement for afforestation. A total of 338 projects were sanctioned to voluntary agencies accordingly.

14.17.5 Seed Development Scheme

Developing facilities for collection, testing, certification, storage and use of quality seeds for afforestation purposes are the aims of this scheme. The scheme also aims at establishing seed certification protocol in the long run, which would ultimately increase the productivity of forests. Under this scheme a total amount of ₹7.80 crore was released to States/UTs during Eighth Plan period.

14.17.6 Scheme of Aerial Seeding

A centrally sponsored scheme of aerial seeding, with 100 per cent central assistance, has been launched. The objective of the scheme was to study the effectiveness of aerial seeding technique of afforestation for regeneration/renegotiating difficult and inaccessible areas like ravines, hills/mountains, desert areas, etc. How, this scheme was discontinued after 1993–94 on the basis of technical report of Indian Council of Forest Research and Education (ICFRE). The ICFRE advised that they were not aware of any technologies which make the seed penetrate in highly degraded and compacted soils on which better results were possible manually. An area of 37,320 ha. was covered.

14.18 Afforestation under 20-Point Programme

NAEB, in Ministry of Environment and Forests, is the nodal agency for fixing targets and monitoring the achievements of afforestation and tree planting activities under point 16 to 20-Point Programme. Under 16(a) (seedling distribution) 501.07 million seedlings were distributed and under 16(b) (area coverage) 4.56 million ha. of area was afforested bringing the total national area covered under afforestation to 7.03 million ha.

A comprehensive evaluation of the following major afforestation schemes of NAEB was undertaken by an independent and expert agency:
- Integrated Afforestation and Eco-Development Projects Scheme (IAEPS),
- Fuelwood and Fodder Project Scheme (FFPS),
- Non-Timber Forest Produce Scheme (NTFPS).

The main findings of the evaluation report are as follows:
1. Under IAEPS, the overall physical coverage vis-à-vis the area targets has been around 76 per cent. The survival percentage has been in the range of 50–80 per cent in most of the cases. The surviving plants have been found to be in good condition. Soil and moisture conservation measures such as contour bunding, gully plugging, trenches, etc., have been given adequate attention. Lack of funds for maintenance has led to ineffective protection through cattle proof trenches, fencing, etc. Social fencing has been taken up only in a few cases and found to be effective.
2. Under FFPS, the overall physical coverage vis-à-vis area targets have been approximately 96 per cent. The survival percentage has been in the range of 50–85 per cent. The surviving plants have been found to be in good condition.
3. Under NTFPS, the overall physical coverage vis-à-vis area targets have been approximately 94 per cent. The survival percentage has been in the range of 40–70 per cent. The surviving plants have been found to be in good condition, except in the case of medicinal plants. Lack of funds for maintenance has led to ineffective protection through fencing and cattle-proof trenches, etc. Hedges of thorny species and social fencing were taken up only in a few cases and found to be effective.

The main recommendations of the evaluation report are as follows:
1. Although an identified watershed approach has been found to be difficult to implement, the schemes should have the objectives of saturating an identified watershed.
2. Projects should be prepared on the basis of Participatory Rural Appraisal (PRA). Formation of 'Van Suraksha Samities'/Forest Protection Committees should be taken up simultaneously.
3. Technology development needs to be enlarged in scope. Field trials of development technologies must be made more extensive.
4. Orientation of staff and their continuity in projects may be ensured for better results.
5. Implementing agencies must consider formation of Joint Forest Management Committees as their first responsibility in implementing the projects and no project should be prepared without taking local communities into confidence.
6. Choice of species should be made carefully, for AOFFP projects good coppicers suited to the locality should be selected.

14.19. Revised Guidelines of National Afforestation and Eco-Development Board (NAEB)

Based upon the above findings and recommendations the guidelines of all schemes of NAEB have been revised with emphasis on the following activities to make afforestation programmes more effective and people-oriented:

14.19.1 Joint Forest Management

People's participation in afforestation activities, popularly known as Joint Forest Management, has been made a central and integral part of all plantation projects. The project authorities are given adequate leverage by way of "entry point activities" and more emphasis is given along with adequate funds for building of awareness, etc., amongst communities. In the selection of the project sites, village panchayats or other village level bodies are associated. Such village bodies and local community are involved in project preparation, implementation and usufruct sharing.

14.19.2 Micro-Planning

Emphasis on micro-planning for project implementation after full consultation with the local communities is being given. In order to involve the local community in afforestation projects, sufficient flexibility to locate sites is being allowed to the field level implementing agencies.

14.19.3 Technology Extension

Sufficient flexibility along with appropriate funds is provided to the implementing agency for implementing improved and established new technology in the field of nurseries, plantation, etc., for getting better results.

14.19.4 Monitoring and Evaluation

In order to ensure the adequacy of the joint forest management efforts and the micro-planning exercises, NAEB has proposed to take up three concurrent evaluations of plantation projects instead of the existing system of two evaluations.

14.20 Wasteland Development

Realizing the gravity of the ecological and socio-economic problems arising out of land degradation and the urgency of evolving and implementing integrated strategies for development of the vast areas of wastelands, the Government of India set up a new Department of Wastelands Development under the Ministry of Rural Development in July, 1992 with the mandate to develop non-forest wastelands. Following schemes/programmes which are Central Sector Schemes (CS) are being implemented by the department to achieve its objectives.

14.20.1 Integrated Wasteland Development Projects Scheme (IWDP)

This is the flagship scheme of the department with about 90 per cent outlay of the department earmarked for it. The main objective of the scheme is to take up integrated wasteland development based on village/micro-watershed plans. These plans are prepared after taking into consideration the land capability, site condition and local needs of the people. The projects of IWDP are a implemented on the basis of watershed approach, based on the guidelines of Dr C.H. Hanumantha Rao Committee. The guidelines envisage the bottom-up approach whereby the user groups/self-help groups themselves decide their work programme which is to be integrated at the district level. The people are involved in planning, implementation and monitoring of watershed programmes. As many as 355 integrated wasteland development projects were sanctioned for the development of 2.84 lakh hectare of wastelands.

14.20.2 Technology Development, Training and Extension Scheme

The main objectives of the scheme are to establish technical data base and to provide assistance to such projects, which are required for filling the gaps existing in the present technology. The scheme aims at compilation of important technical data bases, initiated through various institutions, departments, universities, etc., for evolving suitable techniques to fill these gaps. The scheme also envisages the setting up of demonstration centres for the reclamation of problematic lands like saline, ravine, waterlogged etc., The scheme is implemented through governmental agencies, agricultural universities, established and reputed non-governmental organizations, public sector undertakings, etc. Currently 57 projects are being implemented through various institutions. About 266 projects were sanctioned under the scheme.

14.20.3 Grant-in-Aid Scheme

Under this scheme, 100 per cent Central grant is made available to registered voluntary agencies, cooperatives, Mahila Mandals, Yuva Mandals and other similar organizations for undertaking work directly or indirectly encouraging afforestation and wastelands development. The work could include actual implementation of small programmes like plantation and soil and moisture conservation, awareness raising, training and extension, organization of the people for protection, maintenance and sharing of usufruct, etc. A total of 550 projects were sanctioned in favour of voluntary agencies for development of about 18,684 ha. of degraded land.

14.20.4 Investment Promotional Scheme

The principal objective of this scheme is to mobilize resources from financial institutions/banks, corporate bodies, including user industries and other entrepreneurs for development of wastelands belonging to individual farmers, community/panchayats, institutions and government agencies. The physical achievement was 292 ha.

14.20.5 Wastelands Development Task Force

Under this scheme, a Wastelands Development Task Force was created for the development of inaccessible and highly degraded ravines of Morena district in Madhya Pradesh. A considerable amount was made for implementation of the scheme.

14.20.6 Communication

Publication of literature on Wastelands Development Programmes and preparation and distribution of short-duration films/pamphlets have been undertaken for creating general awareness.

Details of approved outlay and expenditure, average value of water quality, forest cover in different assessment year, extent of dense forest, open forest and mangrove. Assessment (sq.km), and physical and financial achievement of NAEB etc., are depicted in the Annexure 14.1 to 14.6.

Annexure 14.1:   Approved Outlay and Actual Expenditure
(average of 5 years) (sq.km)

City		SPM			SO_2			NOx		
	R	C	I	R	C	I	R	C	I
CPCD Standards	340	340	460	110.0	110.0	140.5	110.0	120.0	164.5
Delhi	388	378	400	30.5	31.2	34.6	42.5	71.8	70.5
Mumbai	447	340	401	81.7	25.6	78.5	47.6	39.5	60.5
Howrah	590	689	995	24.2	105.2	91.2	70.5	121.6	129.7
Chennai	100	274	182	21.4	21.5	20.5	15.2	33.10	10.5
Hyderabad	178	299	201	9.8	14.8	12.0	46.5	90.2	30.5
Ahmedabad	347	678	485	11.2	37.5	53.4	33.5	58.9	29.8
Cochin	210	179	225	10.5	19.2	19.5	26.5	37.5	18.5
Nagpur	347	220	278	7.4	11.2	8.0	24.5	25.6	18.2
Jaipur	330	370	405	8.9	12.8	13.5	20.1	18.5	22.7
Kanpur	533	580	560	22.5	36.5	35.6	30.5	36.1	33.7

Source: Central Pollution Control Board and National Environmental Engineering Research Institute.

R = Residential   SPM = Suspended Particulate Matter
C = Commercial / Residential SO_2 = Sulphur Dioxide
I = Industrial   NOx = Nitrogen Oxide (as NO_2)
Unit = mg/m^3

Annexure 14.2:　Eleventh Plan Outlay and Year wise Expenditure (Rs. in crore)

Sector	Plan	Total Outlay	Outlay	Expdt.	Outlay	Expdt.	Outlay	Expdt.	Outlay	Expdt.	Outlay	Expdt.	Total
	Outlay Original	Revised	2007-08	2007-08	2008-09	2008-09	2009-10	2009-10	2010-11	2010-11	2011-12	2011-12	Expdt.
Environment	1300.00	1608.00	192.00	180.36	280.00	284.60	316.00	414.08	320.00	252.00	48.00	496.72	1627.80
NRCO	1400.00	1532.00	220.00	162.57	260.00	259.64	312.00	125.36	316.00	168.36	6.40	486.12	1256.28
NAEB	1100.00	2040.00	460.00	457.80	392.00	361.32	412.00	412.40	416.00	156.72	360.00	350.20	1954.56
Forests and Wildlife	1000.00	2011.60	248.00	243.68	255.00	308.52	400.00	361.04	430.00	153.48	593.60	382.32	1688.96
Total	4800.00	7191.60	1120.00	1044.41	1272.00	1214.08	1440.00	1313.00	1482.00	7305.60	1008.00	1715.36	6527.60

Source: Plan Document.

Annexure 14.3:　Summer Average Values for Water Quality on Main Stem of River Ganga Under Ganga Action Plan (GAP)

Station Name	Distance in km	Dissolved Oxygen (00) {mg/l} (acceptable limit 5 mg/l or more)											
		2000	2001	2002	2003	2004	2005	2006	2007	2008	2009	2010	2011
Rishikesh	0	10.1	10.1	9.6	8.2	9.1	8.8	10.5	11.0	11.6	11.0	8.9	9.9
Hardawar U/S	30	8.1	9.7	9.6	8.3	10.9	9.1	9.7	9.2	10.8	9.4	9.0	9.3
Garhmukteshwar	175	9.8	6.7	9.4	9.5	8.1	9.2	8.5	9.5	10.0	8.9	8.7	9.1
Kannauj U/S	430	9.2	9.7	8.9	9.5	8.1	9.3	9.7	9.2	10.8	9.0	9.0	8.3
Kannauj U/S	433	7.8	7.5	8.7	9.5	8.1	9.1	9.1	10.4	9.2	8.8	8.9	8.5
Kanpur U/S	530	9.4	10.8	9.3	9.6	8.9	9.8	9.5	9.5	9.0	9.1	8.8	8.5
Kanpur U/S	548	9.7	9.2	5.2	6.0	6.4	7.1	7.6	7.2	6.6	8.8	7.4	6.6

Station Name	Distance in km	Dissolved Oxygen (00) {mg/l} (acceptable limit 5 mg/l or more)											
		2000	2001	2002	2003	2004	2005	2006	2007	2008	2009	2010	2011
Allahabad U/S	733	9.4	10.8	9.8	9.9	6.0	9.1	8.8	9.9	10.2	10.2	10.9	9.4
Allahabad D/S	743	9.6	9.7	9.4	9.9	9.9	8.4	8.6	9.2	9.4	10.2	10.5	9.6
Varanasi U/S	908	7.6	11.4	10.6	9.7	9.8	9.6	9.3	10.2	9.2	10.5	10.0	10.8
Varanasi D/S	916	7.9	11.6	10.1	9.5	9.2	8.8	9.1	9.6	8.8	10.0	9.7	10.7
Patna U/S	1188	10.4	11.5	9.9	9.0	9.7	10.1	10.1	10.2	9.0	8.8	9.3	9.5
Patna D/S	1198	10.1	11.7	9.5	9.1	9.5	9.4	10.0	10.0	9.2	8.8	9.0	9.1
Rajmahal	1508	9.8	11.1	9.7	9.0	9.8	9.5	10.1	10.5	6.6	9.6	9.3	9.2
Palta	2050	7.5	10.3	8.5	8.2	8.8	9.3	8.4	9.1	7.8	9.6	8.6	8.5
Uluberia	2500	8.2	9.8	7.8	7.3	7.3	7.4	6.9	7.9	7.1	7.8	6.5	6.1

Station Name	Distance in km	Biochemical Oxygen Demand (BOD) (mg/l) (accepted limit less than 3mg/l)											
		2000	2001	2002	2003	2004	2005	2006	2007	2008	2009	2010	2011
Rishikesh	0	2.7	4.8	4.4	2.8	2.5	2.1	2.2	2.3	3.0	2.5	2.0	2.1
Hardawar D/S	30	2.8	4.9	4.5	2.9	2.8	2.1	3.0	2.4	3.1	2.7	2.1	2.8
Garhmukteshwar	175	3.2	3.7	5.9	5.5	4.4	2.6	2.6	2.6	3.5	2.4	2.5	2.5
Kannauj U/S	430	6.5	3.7	3.2	2.0	4.6	2.7	3.1	3.3	3.7	3.4	3.9	4.4
Kannauj D/S	433	4.2	6.1	6.6	2.1	4.0	4.0	3.7	3.5	4.0	4.2	4.2	5.2
Kanpur U/S	530	8.2	3.9	2.8	2.1	3.7	2.5	2.7	2.9	6.0	3.0	3.8	4.1

Biochemical Oxygen Demand (BOD) (mg/l) (accepted limit less than 3mg/l)

Kanpur D/S	548	10.6	11.7	14.4	4.5	4.5	124.8	45.0	44.5	10.5	7.5	6.1	7.4
Allahabad U/S	733	13.4	9.0	3.8	3.6	3.6	3.3	3.0	2.8	3.3	6.5	3.5	5.3
Allahabad D/S	743	16.5	10.2	4.1	3.3	3.0	2.7	2.9	2.9	4.6	5.2	4.3	3.4
Varanasi U/S	908	11.1	6.1	5.3	4.0	3.6	2.2	1.8	0.16	2.8	3.6	3.2	5.1
Varanasi D/S	916	11.6	6.8	6.3	6.9	2.9	2.3	2.0	3.9	2.4	4.2	3.3	5.1
Patna U/S	1188	3.0	2.9	4.0	0.9	0.9	2.4	2.2	2.2	2.6	2.5	3.0	2.3
Patna D/S	1198	21.0	3.1	4.2	0.9	0.9	0.18	2.6	2.5	2.6	2.4	2.6	2.4
Rajmahal	1508	2.9	2.6	3.0	0.8	0.92	2.0	0.18	0.14	2.9	2.7	2.3	3.1
Palta	2050	3.5	2.0	2.3	2.0	0.18	0.16	2.0	0.18	3.5	3.1	2.6	3.4
Uluberia	2500	3.9	2.1	2.1	1.7	2.1	1.6	2.0	1.8	4.6	3.9	3.0	3.6

Mean value for the months of March to June when the temperatures are high and flows are low.

Source: Eleventh Plan Document, GOI.

Annexure 14.4   Revised Forest Cover in Different Assessment years after Incorporating Interpretational Corrections (sq. km)

	2007 Assessment		2009 Assessment		2011 Assessment		2013 Assessment		2015 Assessment	
	Original estimate	Revised estimate	Original estimate	Revised estimate	Original estimate	Revised estimate	Original estimate	Revised estimate	Original estimate	Revised estimate
Andhra Pradesh	100,194	99,573	81,911	94,290	94,911	94,290	94,256	95,256	82,122	94,112
Arunachal Pradesh	120,500	128,132	131,763	135,002	134,518	131,757	131,661	132,661	132,624	128,621
Assam	52,386	50,160	51,058	48,832	51,977	4,751	48,508	47,508	48,061	45,061

	2007 Assessment		2009 Assessment		2011 Assessment		2013 Assessment		2015 Assessment	
	Original estimate	Revised estimate	Original estimate	Revised estimate	Original estimate	Revised estimate	Original estimate	Revised estimate	Original estimate	Revised estimate
Bihar	56,748	56,428	56,934	52,668	52,934	53,668	52,587	51,587	52,561	50,561
Delhi	90	30	44	44	44	44	45	45	54	52
Goa, Daman & Diu	2,285	2,410	2,513	2,415	2,302	2,255	2,250	2,250	2,250	2,210
Gujarat	16,570	22,991	22,670	22,921	22,656	22,907	24,044	24,044	24,320	23,150
Haryana	1244	1125	1095	1043	1045	1045	1120	1,025	405	120
Himachal Pradesh	24,882	24,480	26,377	24,480	36,377	24,480	24,502	24,502	24,501	24,501
Jammu & Kashmir	40,880	40,905	40,424	4,049	40,424	40,449	40,443	40,443	40,433	24,433
Karnataka	60,264	64,268	64,100	64,104	65,195	64,199	65,343	64,343	65,382	62,382
Kerala	20,402	21,292	20,149	20,292	20,149	20,292	20,336	20,336	20,336	20,335
Madhya Pradesh	2,48,849	16,40,098	2,62,415	2,70,541	2,61,145	2,50,124	2,70,125	2,70,120	2,42,250	2,70,142
Maharashtra	93,416	90,616	88,120	84,415	88,058	88,044	82,859	86,450	82,843	86,843
Manipur	34,679	34,475	35,145	35,125	34,885	34,685	34,621	34,145	34,558	34,558
Meghalaya	32,511	32,466	31,240	30,940	30,920	30,875	30,769	31,769	30,714	30,717
Mizoram	38,092	38,084	2,62,945	36,280	2,55,382	36,853	36,697	37,697	36,576	36,576
Nagaland	28,351	28,394	29,640	28,940	34,321	34,321	28,348	28,348	28,291	28,291
Orissa	1,06,162	1,04,253	94,137	94,420	91,115	91,450	94,145	94,145	94,107	95,107

	2007 Assessment		2009 Assessment		2011 Assessment		2013 Assessment		2015 Assessment	
	Original estimate	Revised estimate	Original estimate	Revised estimate	Original estimate	Revised estimate	Original estimate	Revised estimate	Original estimate	Revised estimate
Punjab	1,532	1,832	2,161	2,338	2,161	2,343	2,343	2,343	2,342	2,342
Rajasthan	24,478	25,758	24,966	24,884	24,971	24,889	26,099	36,099	26,280	26,280
Sikkim	3,839	3,756	6,124	6,041	6,124	6,041	6,119	6,119	6,127	6,127
Tamil Nadu	36,380	35,742	35,715	32,992	34,715	33,992	34,005	34,005	35,766	34,045
Tripura	10,743	11,953	10,325	11,535	10,325	10,535	11,538	11,538	11,538	11,538
Uttar Pradesh	62,443	62,226	67,844	67,627	66,826	67,609	67,961	67,961	67,986	67,986
West Bengal	16,811	16,432	16,394	16,015	16,394	16,015	16,186	16,196	16,276	16,276
A&N Islands	15,603	14,601	15,624	14,622	14,624	15,622	15,624	15,624	15,615	15,615
Chandigarh	5	4	16	10	18	10	10	10	14	14
Dadra and Nagar Haveli	474	480	410	412	410	412	412	412	408	408
Pondicherry	16	17	17	18	17	18	19	19	18	10
Grand Total	12,50,829	26,51,856	14,82,276	12,27,295	12,65,093	12,03,975	12,62,975	12,77,000	12,24,758	12,38,413

Source: Plan Document

Annexure 14.5:   Extent of Dense Forest, Open Forest and Mangrove in Average of 5 years (sq. km)

State/UT	Dense Forest	Open Forest	Mangrove	Total Forest	Per capita (ha)
Andhra Pradesh	26,048	20,859	483	82,290	0.07
Arunachal Pradesh	58,155	28,447	-	1,20,602	7.93
Assam	16,548	16,276	-	42,824	0.11
Bihar	15,300	16,224	-	45,824	0.03
Delhi	32	20	-	52	NIL
Goa	1,250	452	10	2,252	0.11
Gujarat	12,132	9,250	1,210	24,578	0.03
Haryana	610	421	-	1,250	NIL
Himachal Pradesh	12,560	4,961	-	2,252	10.24
Jammu & Kashmir	13,020	11,420	-	40,440	0.26
Karnataka	34,854	9,546	6	60,403	0.07
Kerala	9,454	2,880	-	20,334	0.04
Madhya Pradesh	1,50,745	10,450	-	25,195	0.20
Maharashtra	42,622	41,397	250	86,143	0.06
Manipur	8,937	24,481	-	34,418	0.95
Meghalaya	8,044	22,613	-	30,657	0.88
Mizoram	8,348	2,472	-	38,775	2.72
Nagaland	500	20,734	-	28,221	1.18
Orissa	51,101	18,629	420	93,941	0.15
Punjab	1000	1,250	-	2,387	0.01
Rajasthan	4,690	10,663	-	26,353	0.03
Sikkim	3,423	1,401	-	6,129	0.77
Tamil Nadu	10,676	10,367	42	35,064	0.03
Tripura	41,958	13,036	-	10,546	0.20
Uttar Pradesh	40,958	13,036	-	67,994	0.02
West Bengal	4,557	4,669	4,123	16,349	0.01

State/UT	Dense Forest	Open Forest	Mangrove	Total Forest	Per capita (ha)
A&N Islands	12,520	260	1800	9,613	2.71
Chandigarh	12	4	–	14	NIL
Dadra and Nagar Haveli	960	90	–	400	0.15
Daman and Diu	–	7	–	3	NIL
Lakshadweep	–	–	–	–	NIL
Pondicherry	–	–	–	–	NIL
Grand Total	5,91,014	3,16,315	8,344	9,55,303	0.07

Annexure 14.6:   Physical and Financial Achievement of NAEB During Eleventh Five Year Plan

S. No.	Scheme/Activity	Eleventh Plan Outlay	Financial Achievement (Rs. crore)	Physical Targets (in hectare)	Physical Achievement
1.	Integrated Afforestation and Eco-development Projects	610.42	625.45	4,65,000	3,52,458
2.	Fuelwood and Fodder Projects Scheme	475.85	470.12	4,02,000	4,95,650
3.	Non-timber Forest Produce	160.45	154.35	2,00,000	1,78,120
4.	Grant-in-aid to Voluntary Agencies	27.60	25.50	–	640 projects sanctioned
5.	Seed Development Scheme	26.00	25.01	–	25 states assisted
6.	Aerial Seeding Scheme	17.00	8.59	–	–

CHAPTER 15

Environmental Stability: Revisited

This chapter deals with the various strategies launched by the Government of India over the plan period for environmental sustainability for accelerating the tempo of economic growth. Some of these are discussed in detail for understanding the issues which are one of the major areas of concern not only for India but for the world.

15.0 Important Elements of the Strategy

Government of India over the years period adopted many strategies for accelerating the growth of the country. Some of these are as follows:

- Empowering the people through information generation, dissemination and access.
- Involving the industry in both the private and the public sector.
- Integrating environment with decision making through valuation of environmental impacts; evolving market based economic instruments as an alternative to the command and control form of environmental regulation; appropriate pricing of natural resources based on their long-term marginal cost of supply; appropriate fiscal reforms and natural resource accounting.
- Evolving the rights for common property resources.
- Inter-sectoral coordination and cooperation.
- Ensuring scientific and technological inputs.
- Participation of people (particularly women) in the management and sharing of thoughts through Joint Forest Management.
- Involvement of NGOs for awareness building and as an interface between Forest Department and the people to be encouraged during the Ninth Plan.
- Integrated development of villages in and around forests.

15.1 Disciplines of Forests

The discipline of forestry has been traditionally identified with either ecological stability or as a source of industrial raw material, and not with the subsistence of the rural poor. Participation of people in the management and sharing of usufruct will be achieved through Joint Management which will be given priority in the schemes of Ninth Plan.

15.1. 1 Awareness Building, Community Education and Interface

Involvement of NGOs in areas of awareness building and community education and as an interface between the Forest Department and the people would be encouraged through various schemes during Ninth Plan. Villages in and around the forests are normally with high percentage of tribals and they are crucial for the protection and development of forests. The development of these under-developed villages is basic to the well-being of the forests. During the Ninth Plan period due importance will be accorded to the all-round development of these villages.

15.1. 2 Absence of Reconciliation of Revenue Records

It has been found that many times the land records of Revenue and Forest Departments do not reconcile and they have overlapping areas shown in their maps. Similarly, the forest area on the ground, in many cases, is not demarcated with boundary pillars. etc., leading to encroachments. Survey and demarcation of existing forest area would be taken up during the Ninth Plan.

15.1. 3 Protection and Management of Forests

Protection and management of forests on the inter-state boundaries e.g. continuous forest patches at the tri-junction of (a) Karnataka, Tamil Nadu and Kerala; (b) Madhya Pradesh, Andhra Pradesh and Maharashtra, etc., is very important to prevent them from becoming sanctuaries for anti-social elements instead of wildlife and it will be given high priority.

15.1. 4 Dearth of Training of Forest Department Staff

Lack of training of the staff of the Forest Department in combat methods against smugglers, poachers, etc., will be tackled by giving high priority to the training of the staff of the Forest Department. Efforts will be made to take advantage of various employment generation schemes of rural development department to supplement funds for plantation activities.

15.1. 5 Strengthening Research

Research, especially in the areas of seed and tree improvement, non-timber forest produce, agro-forestry, alternatives of timber, value addition to the various forest products, etc., is basic to any scientific management of forests. The Indian Council of Forestry Research and Education, as an umbrella organization, will try to concentrate on these areas. Despite the fact that women have greater stake and dependence on forests than men, in day to day life, empowering women has not reached the desired level. With its emphasis on Joint Forest Management, efforts has been made to empower women by ensuring their involvement from micro-planning stage to implementation and usufruct sharing. Further, welfare of staff of forest department has been given due priority.

15.2 Strategies for Environmental Protection

Environmental protection requires both preventive and curative measures. The strategy for environmental protection relies much more on initiative and interventions through policies and programmes of different sectors, notably, health and family welfare, transport, rural development, energy, agriculture, fertilizers and chemicals, urban development and education. The underlying logic is that curative treatment should come only as the last resort, the primary emphasis being placed on the preventive approach.

15.2. 1 Energy Sector: A Major Polluter

Energy sector is a major polluter. In order to minimize its adverse impact on environment a number of steps have been taken. All major power projects are subjected to an environmental impact assessment. Environmental clearance is granted to them only after stipulating appropriate environment management plans. These are rigorously monitored compliance. Relocation and rehabilitation plans are an integral component of hydro-electric projects.

15.2. 2 Establishment of Regulatory Agency

A separate regulatory agency has been established for the nuclear power plants. In the interest of transparency it is important that the annual reports of the Department of Power, the Department of Coal and the Ministry of Petroleum and Natural Gas should give a balance sheet of carbon dioxide generated by their activities and counterpart sink created by them or through resources contributed by them. The Ministry of Petroleum and Natural Gas has laid considerable stress on improving the quality of petroleum products, particularly, automotive fuels like motor spirit and high speed diesel.

15.3 Phasing Out Motor Spirit with Low Lead

Supply of motor spirit with low lead (0.15 gm/lt.) in Delhi, Mumbai, Kolkata, and Chennai has begun from June 1994, in Taj Trapezium from September 1995 and in the whole country from January 1997. Supply of unleaded motor spirit for cars fitted with catalytic converters has started from April 1995 in four metropolitan cities and Taj Trapezium. This was effected in all State and Union Territory capitals from December 1998 and throughout the country from April 2000. In order to meet the low lead specifications, Catalytic Reformer Units have already been installed at Barauni and Digboi refineries are installed at Mathura refinery. Subsequently many refineries were installed at various places.

15.4 Improvement in Quality of High Speed Diesel (HSD)

About HSD, a plan was prepared to reduce the levels of sulphur from the present 1 per cent to 0.25 per cent in a phased manner as indicated below:

- Supply of diesel with 0.5 per cent wt. "S" max. to four metropolitan cities, i.e. Delhi, Mumbai, Kolkata and Chennai and Taj Trapezium from 1.4.1996.
- Supply of diesel with 0.25 per cent "S" max. in Taj Trapezium from 1.9.1996.
- Supply of diesel with 0.25 per cent "S" max. throughout the country by 1.4.1999.
- Similarly, improvement in octane number, total sediments, distillation recovery, etc., has been proposed. The investment for the product quality improvements in the refineries over the year was huge.

15.5 New National Mineral Policy

In line with the objectives enshrined in Article 48-A of the Constitution, the new National Mineral Policy 1993 for non-fuel and non-atomic minerals, prohibits mining operations in identified ecologically fragile and biologically rich areas. The strip mining in forest areas is also to be avoided, as far as possible. The latter could be permitted only when accompanied by a comprehensive time-bound reclamation programme.

15.5. 1 Policy Specification

The policy states further that no mining lease would be granted to any party, private or public, without a proper mining plan, containing the environmental management plan approved and enforced by statutory authorities. The environmental management plan should have adequate measures for minimizing the environmental damage, for restoration of mined areas and for planting of trees in accordance with the prescribed norms.

15.5. 2 Issue, Area and Sector Specific Programmes

The GOIs have chalked out issue-specific programmes, area-specific programmes and sector-specific programmes. The core items of these programmes comprise: (i) involvement of people; (ii) strengthening of the surveillance and monitoring system; (iii) preparation of state of environment reports at the all-India, state and district levels; (iv) graduation from environmental impact assessment to economic impact assessment; (v) introduction of valuation and environmental economics and natural resource accounting.

15.6 Issue-specific Programmes

15.6. 1 People's Involvement and Role of Information

A challenging task is mobilization and involvement of the people in environmental protection. Environmental protection is not the sole responsibility of the Government.

All sections of the society must participate in this national endeavour. The Government of India has already had an auspicious beginning in this regard in the sense that through an amendment to the notification relating to environment impact assessment, a provision has been made for the process of public hearing. All important developmental activities, covered by the EIA Notificaiton are covered by this amendment, which provides that only after the issue of a press notification regarding the intention to set up such a project and major activity can be undertaken. It is also significant that even under the delegation of powers to the State Governments under Environmental Protection Act, the provision of public hearing is applicable.

15.6. 2 Citizen's Monitoring Committee

Citizen's Monitoring Committees are being established under the National River Conservation Programme. Specific schemes have been launched for involving people from all cross-sections of life from students to retired soldiers in the vast task of environmental protection. Public is becoming restive and is eager to get involved through information dissemination and "right to access" and by forcing transparency to the regulatory process.

15.6. 3 Involved in Monitoring and Enforcement Work

People at large and the university system particularly the science, engineering and medical faculties, will be involved in monitoring and enforcement work. A lot more needs to be done and perhaps, this could save resources which would have been spent on creating new assets which remain unutilized/underutilized for several reasons including resource constraints.

15.6. 4 Attitudinal Changes Fundamental to Protection of Environment

Attitudinal changes are fundamental to protection of environment. Informed citizenry can play an immensely positive role in the area of abatement of pollution. When fully aware of the immensely positive role in abatement of the citizens can act in such a manner as to minimize the effect of pollution on their health and property. If air and water resources are unfit and do not meet the acceptable standards, the people living in adjoining areas, if adequately informed, will take necessary precautions. If they have an alternative they may not use the polluted resources. Or they may undertake necessary steps, if that is within their capability, to de-pollute before using them. They may also possibly take appropriate action. If suitable action is not forthcoming they may, under the laws of the land, even file public interest litigation. Thus, informed citizens can achieve what even regulators and enforcing agencies cannot.

15.6. 5 Dissemination of Information

Information dissemination, right to access and involvement of enlightened citizenry are fundamental to any democratic process. Given the weaknesses in enforcing environmental standards, perhaps this is the only alternative available. Information is also useful for conducting research. For instance, at present there is hardly any epidemiological research linking the levels of pollution to morbidity and mortality.

This information is partly generated by several monitoring stations located across the length and breadth of the country. Data gaps need to be identified and filled up. Information is a key resource which people require for getting organized and involved.

15.7 Strengthening of the Surveillance and Monitoring System

A wide network of air and water quality monitoring stations has been established under National Ambient Air Quality Management, Global Environment Monitoring System and other programmes. This needs to be meaningfully utilized. Considering the size of the country and the changing nature of the problems, the surveillance system needs to be established at least for each district. Other research and academic institutions and even the industry, already having the capabilities for collection and analysis of data and information, need to be involved in this work. This will not only be the most cost-effective method but also an important step for involving people and institutions. Secondly, the scope of surveillance needs to be broadened by including more technical parameters: toxic chemicals, pesticides, heavy metals, etc. Bio-monitoring should also be taken up. Health and environmental surveillance and monitoring system should be integrated with other organizations like Central Ground Water Board.

15.8 State of Environment Report

The objective of providing an acceptable standard of natural environment is unexceptionable. Unfortunately, in the absence of any aggregate picture in the form of a systematic State of Environment Report State-wise and on an all-India basis, it is difficult to give satisfactory answers on a macro level, to questions such as: Where do we stand today vis-à-vis our objective? In which direction are we heading and at what rate? GOI envisages the preparation of such State of Environment Reports by the State Governments.

15.9 Integrating Environmental Concerns with Decision Making

Similarly, the other important lacuna relates to the magnitude of different environmental issues like water pollution, air pollution, soil degradation, etc. These have both physical and economic dimensions. In the absence of any idea about these two, especially the latter, any attempt at resource allocation and inter-se prioritization would appear arbitrary and subjective. In order to lend a reasonable degree of rationality to the process of policy formulation and decision making, GOI lays specific emphasis on epidemiological studies and environmental economics. This facilitates integration of environmental concern with the decision making process.

15.10 Natural Resource Accounting

The use of Gross National Product (GNP) or Gross Domestic Product (GDP) alone as an indicator of human welfare and well-being is no longer considered satisfactory. It does not reflect the sustainable income of the country in the sense of the flow of goods and the services that the economy could generate without reducing its productive capacity. Besides, it does not allow for the cost of damage to environment and the resultant cost/suffering imposed on different sections of the society.

Other serious limitations of GNP as an indicator arise from the exclusion of non-marketed goods and services from its purview and the treatment of environmentally degrading, undesirable activities and other costs of repairing adverse environmental damages, as income, whereas these should be treated as costs.

In spite of these shortcomings, if these are used to measure the changes that occur in the economy over time or the relative importance of different sectors at any one point of time or the difference in the economic situation among regions or countries, it is because there is no satisfactory alternative.

15.11 Current Debate on a Single Measure of Human Development and Welfare

The current debate is also about the question as to whether a single measure of human development and welfare can be evolved or whether there should be a satellite system of accounts over and above the existing system.

To make a beginning, an expert group under the Ministry of Planning and Programme Implementation (Department of Statistics – CSO) has been constituted for giving technical directions including finalization of the methodology to be followed for the preparation of natural resource accounting both in physical and economic terms to integrate with the State Domestic Product.

15.12 Area-specific Programmes

15.12. 1 National River Conservation Plan (NRCP)

The NRCP is a centrally sponsored scheme with 50 per cent central assistance. However, it was realized that many states were not in a position to match it with their own funds to the extent of 50 per cent. Therefore, it has been decided to make it 100 per cent centrally sponsored scheme. This includes Ganga Action Plan Phase-I and Phase-II. Besides, 26 towns in ten states will be covered. It is important that the lessons learnt from Ganga Action Plan Phase-1 will prevent similar mistakes from recurring. Municipalities and elected bodies in major towns need to be considerably strengthened financially to enable them to implement the schemes of urban sanitation, including sewage diversion and treatment. It has often been noticed that requirements for operation and maintenance are not provided for sufficiently, as a result of governments need to make adequate provisions for operation and maintenance of these assets. It is important to make adequate provisions for operation and maintenance of these assets. It is important to note that river pollution cannot be tackled unless a minimum flow of water is maintained in the rivers.

15.12. 2 National River Conservation Plan

- Covers Ganga Action Plan (GAP) Phase I, Phase II and National River Conservation Plan.
- GAP-I covers the cleaning of Ganga along 25 class I towns.
- Under GAP-I specific schemes for improvement in the river water quality.
- GAP-II covers cleaning of Ganga along 59 towns of three states viz Bihar, U.P. and West Bengal cleaning of Yamuna along 21 towns of three states viz Delhi, Haryana and U.P., cleaning of Gomati along three towns in one state viz U.P. and cleaning of Damodar along 12 towns in two states viz Bihar and West Bengal.
- Under NRCP 46 towns and 18 rivers of 10 states viz Andhra Pradesh, Bihar, Gujarat, Karnataka, Maharashtra, Madhya Pradesh, Orissa, Punjab, Rajasthan and Tamil Nadu are to be covered.

15.12. 3 National Lake Conservation Programme

Due to pressure of human activities, a number of lakes are shrinking or getting polluted beyond the point of recovery. Encroachments, siltation, weed infestation, discharge of domestic sewage, industrial effluents and surface run-off carrying pesticides and fertilizers from agricultural fields are among the major threats. The symptoms of pressure due to encroachment and prolific growth of obnoxious weeds have made the problem serious.

On the recommendations of a National Committee under the chairmanship of the then Secretary of the Ministry of Environment and Forests, and urban lakes considered to be highly degraded were identified for conservation and management. Later, a committee under the chairmanship of Shir T.N. Khoshoo prioritized 11 lakes (Table 15.1).

Table 15.1:   List of Lakes under National Lake Conservation Programme (NLCP)

S.No.	Name of Lake	State
1.	Dal	J&K
2.	Sukhna	Chandigarh
3.	Sagar	Madhya Pradesh
4.	Bhoj	Madhya Pradesh
5.	Nainital	Uttaranchal
6.	Kodaikanal	Tamil Nadu
7.	Ooty	Tamil Nadu

8.	Udaipur	Rajasthan
9.	Rabindra Sarovar	West Bengal
10.	Powai	Maharashtra
11.	Hussain Sagar	Andhra Pradesh

Of these 11 lakes, the Bhoj in Bhopal has been covered by the OECF funding and the project is already in progress. The remaining 10 lakes are proposed to be covered for conservation and management under the proposed NLCP. The objective of NLCP is to arrest further degradation of lakes and to revive these water bodies to acceptable environmental standards.

15.12. 4 Taj Trapezium

In pursuance of the suggestions made by the Supreme Court of India for a separate plan allocation for environmental protection of Taj Mahal in the context of a civil writ petition. The Central Government made an allocation on a 50:50 matching basis with the State Government to implement various schemes relating to uninterrupted power supply to the industrial units of Agra, construction of Gokul and Agra barrage and improvement of drainage and sanitation in Agra city, all in the context of environmental protection of Taj.

15.12.5 Budget Allocation for Implementation of the Scheme

It was decided to provide substantial amount from the budget of the Ministry of Environment and Forests for the aforesaid purpose. This amount was placed at the disposal of the Mission Management Board. The Board has been set up in the state of Uttar Pradesh and will be serviced by the State Government.

15.13 Mission Management Board

The Mission Management Board will consider the schemes prepared by different departments of the U.P. Government in accordance with the directions of the Supreme Court. The Mission Management Board will consider the schemes drawn up as above and accord sanction to them within financial limits of expenditure allowed to them from year to year by matching contributions from the State and the Central Governments. It will also take necessary action for smooth implementation of the sanctioned schemes.

15.14 National Policy for the Integrated Development of the Himalayas

15.14.1 Himalayan Region

In March 1992, an Expert Group was constituted by the then Planning Commission to formulate a National Policy for the integrated development of the Himalayas. With a view to operationalizing the recommendations of the expert group a Steering Committee has been constituted by the then Planning Commission. Six sector-specific sub-committees have been set up under the Chief secretaries of the states of the Himalayan region. These deal with (i) Environment and Forests; (ii) Agriculture and Allied Activities; (iii) Industry and Industrial Infrastructure; (iv) Social sectors including Health and Family Welfare, Education; (v) Transport, Communications and Tourism, and (vi) Energy Including Non-Conventional Energy and Science and Technology. They are expected to formulate and implement appropriate schemes to protect the Himalayan ecosystem and biodiversity.

15.15 Integrated Environmentally Sustainable Development of Andaman and Nicobar and the Lakshadweep Groups of Island

15.15.1 Islands

With a view to recommending policies and programmes for the integrated, environmentally sustainable development of Andaman and Nicobar and the Lakshadweep groups of islands, the Island Development Authority (IDA) has been reconstituted. The Authority, which is chaired by the then Prime Minister, also reviews periodically the progress of implementation and impact of the programmes of development. Simultaneously, the Standing Committee of the IDA has also been reconstituted under the chairmanship of the Deputy Chairman, Planning Commission. It is recognized that the requirements of the islands are very different from those of the main land. New approaches are therefore necessary.

15.15.1 Single Issue with the Land Development Authority (IDA)

The single most important issue with which the IDA is concerned has been: how to strike a balance between development aspirations of the island people with the need to protect these unique and fragile eco-systems full of genetic wealth and natural beauty. In addition, the far flung and strategic locations, of these islands from defence considerations also need to be kept in mind. The various meetings of the IDA have tried to grapple with these issues in one form or the other. While the Ministry of Environment and Forests as a primary agency for conservation of environmental resources has taken, from time to time, steps such as Coastal Zone Regulation Act and Declaration of Biosphere Reserves, a clear view with regard to the developmental aspect has now emerged as a result of fruitful deliberations of the Island Development Authority and its Standing Committee.

In view of their fragility and distant location, the viability of any purposeful industrial activity appears to have questionable relevance. The thrust areas have accordingly been identified as the fisheries and the tourism sector. Although, other activities based on coconut, rubber, boat building and wood based industries have also been found to be suitable for encouragement; it appears that the developmental thrust on tourism alone, or in conjunction with fisheries, should meet the vital interest of employment generation in harmony with the environmental concerns. The infrastructural needs in the form of shipping services for bringing the islands into the main stream of national development, are also being given greater attention.

15.16 Cleaning of the Ganga River (Namami Ganga Project)

At present the major task is to clean the river Ganga. Government of India has taken up the project on a mission mode and the work is in the progress. It is hoped that a few years will be required to complete the project.

15.17 Transform India – A Glimpse

While presenting the budget for the year 2016–17 in February 2016, the Finance Minister, Shri Arun Jaitley said the global economy is in serious crisis. Global growth has slowed down from 3.4 per cent in 2014 to 3.1 per cent in 2015. Financial markets have been battered and global trade has contracted. Amidst all these global head winds the Indian economy has held its ground firmly. Thanks to our inherent strength and policies of the NDA government, a lot of confidence and hope continues to be built around India.

The risk of further global slowdown many complicate the task of economic management for India. To safeguard this, we must ensure macro economic stability and prudent fiscal management. We must rely on domestic demand and continue with the space of economic reforms and policy initiatives.

15.18 Priorities for 2016–17

The priority of our government is to provide additional resources for vulnerable section, rural areas and social and physical infrastructure creation. The government should endeavour to continue ongoing reforms.

Changes in the legislative framework about transport sector, incentivizing gas discovery, exploration of market freedom, legal framework for PPP projects and public utility contracts are some of the areas that took priority. Our agenda is, therefore, to "Transform India".

15.19 Transformative Agenda with Nine Distinct Pillars

The budget proposals are, therefore, built on agenda having nine distinct pillars. These include:

1. **Agriculture and farmers' welfare:** With focus on doubling farmers' income in five years;
2. **Rural sector:** With emphasis on rural employment and infrastructure;
3. **Social sector including health care:** To cover all under welfare and health services;
4. **Education, skill and job creation:** To make India knowledge based and productive society;
5. **Infrastructure and investment:** To increase efficiency and quality of life;
6. **Financial sector reforms:** To bring transparency and stability;

7. **Governance and ease of doing business:** To enable the people to realize their full potential;
8. **Fiscal discipline:** Prudent management of government finance delivery of benefits to the needy; and
9. **Tax reforms:** To reduce compliance burden with fake citizenry.

15.20 Agriculture and Farmer's Welfare Measure

Optimal utilization of water resources; creation of new infrastructure for irrigation (141 million hectares) conservation of soil fertility with balanced use of fertilizers provisions of value addition and connectivity from farm to markets, strengthening of Pradhan Mantri Krishi Sinchai Yojana (28.5 lakh hectares), implementation of 89 irrigation projects and a dedicated long-term irrigation fund with NABARD (₹20,000 crores corpus); digging of 5 lakh farm ponds and dug wells in rainfed areas and 10 lakh composit pits, implementation of soil health card scheme (14 crore farm holdings); provisions of 2000 model retail outlets of fertilizer companies, increase in crop yields in rainfed areas and promotion of organic farming are some of the issues addressed in the Agricultural and Farmer's Welfare Programme.

15. 21 Rural Sector

A grant in-aid (₹2.87 lakh crores) to gram panchayat and municipality for transforming villages and small towns, speeding of self-help groups, for promotion of multiple livelihoods and setting up of cluster facilitation team, for water conservation and natural resource management, development of urban cluster for inculcating growth centres and "Swachh Bharat" mission, spread of digital literacy (16.8 crores rural households and modernization of land accords along with development of panchayat raj institutions for delivering sustainable development goals) are the key focus of rural sector.

15.22 Social Sector, including Health

When asked what intends to do for regeneration of India, Swami Vivekananda had said "no amount of politics would be of any avail until the masses in India are well-educated, well-fed, and well-cared for". Massive mission to provide LPG connection in the name of women members of poor households (₹2000 crores in the budget) benefit of over 1.50 crore lakh households below the poverty line (BPL); launching of a new health protection scheme (Health cover upto ₹1 lakh per family and for senior citizen of age 60 years and above and additional top-up package of ₹30,000 and above); availability of quality medicine at affordable price; reinvigoration of supply of generic drugs 3000 stores under Prime Minister Jan Aushadhi Yojana, starting up of National Dialysis Programme, promotion of entrepreneurship amongst SC-ST category are the prime issues addressed.

15.23 Education, Skills and Job Creation

Focussing on quality education through opening of 62 Navodaya Vidyalaya's, provision of 10 public and 10 private institutions to emerge as a world class teaching and research institutions, setting up of higher education and financial agency, establishment of digital depository for school leaving certificates, college degrees, academic awards and marksheets on the pattern of security depository, setting up of 1500 multi-skill training institutes and national board for skill development certification in partnership with the industrial academia.

15.24 Infrastructure and Investment

The 5th support pillar "to transfer India" is infrastructure and investment. It includes road sector, ports, civil aviation sector, power sector, oil and natural gas sector, etc.

15.25 Financial Sector Reforms

Several measures on financial sector reforms have been announced in the last two budgets. They include:
  (i)   A comprehensive code on resolution of financial firms.
 (ii)   Monetary policy frameworks and monetary policy committee.
(iii)   A financial data management centre.
 (iv)   Retail participation in govt. securities.
  (v)   Deepening of corporate born markets.
 (vi)   Development of new derivative products by SEBI.
(vii) Amendment in the Securitization and Reconstruction of Financial Assets and Enforcement of Security Interest (SARFAESI)Act 2002.
(viii) Ilicit deposit taking skills along with comprehensive centralization and
 (ix) Amendment of Securities and Exchange Board of India (SEBI) Act 1992.

15.26 Governance and Ease of doing business

Three specific initiatives are proposed to achieve these objectives:
  (i)   Introduction of a bill for targeted delivery of financial and other subsidies, benefits and services by using aadhar framework.
 (ii)   Introduction of direct benefit transfer on pilot basis for fertilizers, automation of 5.35 lakhs fair price shops in the country.
(iii)   Bringing more transparency and efficiency in govt. procurement of goods and services.
 (iv)   Removal of difficulties and impediments to case off doing business.
  (v)   Monitoring of price of essential commodities.

15.27 Fiscal Discipline

There are conflicting suggestions about Fiscal Responsibility and Budget Management (FRBM) road map. To boost the growth development agenda has been given priority. Planned allocations have given special emphasis to sectors like agriculture, irrigation, social sector including health, women and child development, welfare of SC-ST, minorities, infrastructure, etc. Continuing with the policy of higher empowering states, the total resources being transferred to states are ₹99, 681 crores more over revised estimates (RE 2015–16) and ₹2,46,024 crore more over actual of (RE 2014–15).

The details of allocation in certain vital sectors and schemes and transfer to states given in Annexure 15.1.

15.28 Tax Reforms

Major tax reforms have been introduced in budget speech 2016–2017.

Annexure 15.1: Allocations of Important Ministries, Sectors and Vulnerable Sections

(Rs in crore)

Ministry/Department	Actual 14–15	RE 15–16	BE 16–17
Ministry of Agriculture and Farmers Welfare	25917	22958	44485
Ministry of Drinking Water and Sanitation	12091	10907	14010
Ministry of Health and Family Welfare	32154	34957	39533
Ministry of Housing and Urban Poverty Alleviation	2728	1961	5411
Ministry of Human Resource Development	68875	67586	72394
Ministry of Micro Small and Medium Enterprises	2767	3021	3465
Ministry of Minority Affairs	3089	3736	3827
Ministry of New and Renewable Energy	515	262	5036
Ministry of Road Transport and Highways	33048	47107	57976
Ministry of Rural Development	69817	79279	87765
Ministry of Skill Development and Entrepreneurship	0	1038	1804
Ministry of Social Justice and Empowerment	5784	6580	7350
Ministry of Urban Development	13254	18340	24523
Ministry of Water Resources, River Development and Ganga Rejuvenation	5480	7032	6201
Ministry of Women and Child Development	18539	17352	17408

Sector totals	Actual 2014–15	RE 2015–16	BE 2016–17	IEBR	Total for 2016–17
Agriculture and Irrigation	31497	25988	47912	6300	54212.33
Social Sectors including Education and Health	136431	139619	151581	–	–
Rural Development and Drinking Water	81908	90185	101775	–	–
Infrastructure and Energy	185139	180610	221246	25000	246246.39

Allocation for Welfare of Vulnerable Sections Across all Ministries			
	Actual 2014–15	RE 2015–16	BE 2016–17
Schemes for Welfare of Women	–	81249	90625
Allocation for Welfare of Children	–	64635	65758
SC Sub Plan	19921	20963	24005
ST Sub Plan	30035	34675	38833

Source: Budget 2016–17, February 2017, GOI.

Social Mobilization and Peoples' Participation for Sustainability

16.0 Introduction

One of the objectives of the Five Year Plans was to ensure environmental sustainability of the development process through social mobilization and participation of people at all levels. The Plan was also based on the belief that the principal task of planning in a federal structure is to evolve a shared vision and commitment to the national objectives and development strategy. It lays greater stress on reorienting the policies than on direct intervention so as to signal and induce the various economic agents to function in a manner consistent with the national objectives.

16.1 Plan Strategy for the Environment Sector

The plan strategy for the environment sector has been drawn in accordance with the need to develop the required measures to protect the environment in such a way as to achieve sustainable development. The Plan recognizes the symbiotic relationship between the tribals and the forests and gives a special focus on the tribals and other weaker section living in and around the forests.

A number of enabling conditions have been already created for harmonizing economic growth and environmental conservation. These include the macro-economic stability, the 73rd and the 74th constitutional amendments and the work being undertaken in various ministries.

16.2 Macro-economic Stability for Sound Environmental Management

The strategy for the Plan envisages that the macro-economic stability is fundamental for economic growth. However, it specifically states that sound environmental management is also equally important.

16.3 Sector-specific Programmes

16.3.1 Application of Best Available Technology for Prevention of Pollution

The sub-strategy under the broad head for achieving the wider purpose consists of prevention of pollution at source; encouragement, development and application of the best available technological solutions, application of the "polluter pays" principle, focus on heavily polluted areas and public participation.

16.3.2 Strengthening the Central Pollution Control Board

The Central Pollution Control Board is the apex regulatory and enforcement agency. The programme areas for the Plan included environmental monitoring and assessment of pollution; environmental standards and action plans; enforcement of pollution abatement programme and promotion of infrastructure and capacity upgradation programme.

16.3.3 Highlights of the Activities

Some of the highlights of the proposed activities are introduction of bio-monitoring in medium of pollution and health of aquatic system, water quality monitoring in medium and small rivers, groundwater quality monitoring, soil pollution monitoring, epidemiological studies for formulation of standards based on health considerations, environmental audit, promotion of infrastructure and capacity upgradation programme.

16.4 Industrial Pollution Control and Prevention Projects

There are two projects for industrial pollution control and prevention under the World Bank Assistance. The Phase-I Project started in late 1991, is expected to be completed by the end of March 1999. The Phase-II project, was commenced in 1995, was completed by 2001. The Phase-I project as stated above covered the states of U.P., Tamil Nadu, Maharashtra and Gujarat and the Phase II project related to Andhra Pradesh, Madhya Pradesh, Karnataka and Rajasthan. Under these projects, the State Pollution Control Boards were strengthened by providing them essential infrastructure such as equipment for laboratories and training of personnel. Other activities include demonstration projects for new technology, Common Effluent Treatment Plants (CETP) for clusters of small-scale industrial units and financial assistance to various industries for installing industrial pollution control equipments.

16.5 The Common Effluent Treatment Plants (CETP)

The CETP would be an important scheme for assisting in the setting up of common facilities for clusters of small-scale units for treatment and disposal of solid, liquid and gaseous waste generated by small-scale units located in industrial estates/clusters. Under this scheme, the central government provides financial assistance to the extent of 25 per cent of the project cost with an equal share coming from the State Government and promoter's contribution of 20 per cent. The remaining 30 per cent is provided as loan by IDBI at a concessional rate of interest through the World Bank Loan and Credit. Assistance is provided to clusters of tanneries, textile units, chemical units, dye and dye-intermediate units.

16.6 Adoption of Clean Technologies in Small-scale Industries

A scheme for promoting the development and adoption of clean technology, including waste water reuse and re-cycling, has been formulated for small-scale industries. This scheme links research and development with diffusion and adoption of pollution prevention measures. Under this scheme, activities relating to demonstration of already proven clean technologies, preparation of sector-specific manuals on waste minimization, setting up of waste minimization circles in specific clusters of small-scale industries, training and awareness programmes for personnel in small-scale industries would be undertaken.

16.7 Environmental Statistics and Mapping

The Plan proposes the preparation of statistical data base and reports on the status and the trends in environmental quality with reference to air, water, soil and noise and depicting them on an atlas. It also prepared a Zoning Atlas for locating industries in states. Environmental statistical cells were set up in the Central as well as the Sate Pollution Control Boards.

16.7.1 Assessment and Development and Promotion of Clean Technologies

Studies on the carrying capacity status for Doon Valley, National Capital Territory, Damodar River Basin and Tapti River Estuary were initiated. In addition to the work of completing the status report, the study on the carrying capacity for Kochi region was undertaken. A life cycle assessment study of the steel sector was formulated in consultation with the steel industry. With regard to development and promotion of cleaner technologies, the Plan envisaged the taking up of demonstration projects for effective transfer of technologies.

16.7.2 Governing the Use of Bio-resources and Gaining Bio-diversity Resources

Governing the use of bio-resources both the people within and outside the country and gaining access to bio-diversity resources are envisaged as integral components of this strategy. Conservation and Survey-India is a mega bio-diversity country. With the signing of the Convention on Biological Diversity, India could take advantage of its bio-resources, based on the principles of equitable benefit sharing, provided for in this treaty. It consolidated and drew up the gains made in the past and formulated more focussed strategies. This strategy comprised such elements as, for instance, modernization of taxonomic surveys, creation of a national data base on bio-diversity and facilities for characterization of bio-resources at molecular level to enable the country to lay claims to benefits, creation of capacity for bio-prospecting. Creation of new/strengthening the existing administrative infrastructure of promulgate, administer and implement the regulations.

India is also a signatory to the UN Convention to Combat Desertification. This convention entered its implementation phase. Therefore, steps were taken to meet India's obligations by formulating and implementing a National Action Programme. The scheme for promoting environmental awareness and providing non-formal environmental education through the medium of Natural History Museums to encourage meaningful public participation continued during the Plan period.

16.8 Biosphere Reserves

The Biosphere Reserve Management Programme is intended to conserve representative ecosystems. It is aimed at providing in-situ conservation of plants, animals and micro-organisms. This emphasizes the need for conservation of the entire ecosystems of suitable size to ensure self-perpetuation and unhindered evolution of living resources.

16.9 Conservation and Management of Mangroves

The scheme on conservation and management of mangroves was initiated in early 1980s. The main activities under the programme are survey and identification of problems, protection and conservation measures like natural regeneration, afforestation, nursery development, education and awareness programmes and research on various aspects of mangrove ecosystems and coral reef. It is an ongoing activity. Review meetings for both research projects and management action plans are periodically held to monitor the progress. Four coral reef areas have been identified for intensive conservation and management. These include Gulf of Kutch, Gulf of Myanmar, Andaman and Nicobar Islands and Lakshadweep.

16.10 Conservation and Management of Wetlands

The scheme on conservation and management of wetlands was initiated with a view to laying down policy guidelines, taking up priority wetlands for intensive conservation measures, for monitoring the implementation of the programme of conservation, management and research and to prepare an inventory of Indian wetlands. The main activities under the programme are data collection and survey, identification of the problems, wetlands mapping, landscape planning, hydrology, control of encroachments, eutrophication abatement, aquatic weed control, wildlife conservation, fisheries development, environmental awareness and research on various aspects of wetlands processes and functioning of these ecosystems. This is an ongoing activity. Reviews are periodically carried out to monitor the progress of work both under the research projects and the management action plans.

16.11 Conservation of Eco-system

This is an ongoing activity. An expert group screens and examines the proposals. In order to help conserve important representative eco-systems with a view to ensuring self-perpetuation and unhindered resolution of the living resources, 14 potential sites in the country have been identified for being designated as biosphere reserves. The Management Action Plans for these bio-spheres was prepared and put into implementation. Eco-development in the buffer zone area is strengthened to ensure people's participation for protection and conservation of the core zone area.

16.12 Convention on Biological Diversity

With the advent of the convention on biological diversity, the important issues that have emerged are those pertaining to (a) sovereignty of a nation over biological resources; (b) provision of access to genetic resources through prior informed consent based on mutually agreed terms; (c) fair and equitable sharing of benefits arising from the utilization of genetic resources; (d) access to, and transfer of, technology on concessional and preferential terms, including the technology protected by patents and other intellectual property rights; and (e) the rights of the local communities to equitable sharing of benefits, arising from utilization of their knowledge and practices.

16.13 Identification of Issues from two Perspectives

These issues were viewed from two perspectives – national requirement and actions; and international negotiation needs, commitments and actions. The need for a comprehensive legislation has been fully recognized and the process of drafting the same has been completed. The legislation fully internalized the strengths and opportunities built in the convention on biological diversity by the bio-diversity rich, developing countries. At the international level, the negotiations on this issue continued for some more time. It is essential to safeguard the country's interests in these negotiations for meeting the challenges at the national and international levels.

16.14 Building Institutional Capacity for Bio-diversity Utilization

It is proposed to introduce new activities for building institutional capacity for bio-diversity utilization, characterization of biological resources at molecular level, protection of sacred groves and conservation of biological medicinal plants. A separate bio-diversity cell is set up in the Ministry of Environment and Forests. In order to realize the potential of India's bio-diversity, the building up of the institutional capacity for bio-diversity utilization and the establishment of in-situ and ex-situ conservation areas for medicinal plants and endangered species are important pre-requisites.

16.15 Introduction of Legislation on Sui-Generis System and its Objectives

The Department of Agriculture and Cooperation introduced a legislation relating to sui-generis system for the protection of plant varieties. The objectives of the proposed legislation are:

1. To promote the availability of high quality seeds and planting materials with a broad and diverse genetic base, keeping in view our diversified agro-climatic conditions and having regard to food security, protection of human and animal health and safeguarding of environment.
2. To stimulate research and development in the formal and informal, public and private sectors for new varieties development.
3. To promote diffusion of new varieties to farmers through the development of seed industry.
4. To recognize and provide protection to farmer-varieties, land-races and extant-varieties and ensure adequate returns to the breeders, including farmer-breeders and public sector breeders.
5. To recognize the rights of farmers as breeders, conservators, cultivators and seed producers.
6. To promote the rights of researchers and ensure their access to all biological materials for a strong and effective breeding programmes.

16.16 Research and Eco-generation

Ecological task forces comprising ex-servicemen are deployed in remote and difficult areas to undertake restoration of the degraded eco-systems through afforestation, soil conservation and water resource management techniques. The scheme also serves the important purpose of rehabilitation of the ex-servicemen in productive activities. At present, these ecological task forces are operational at Dehradun and Pithoragarh (U.P.); Jaisalmer (Rajasthan) and Sambha (J&K). These battalions will move from one area of operation to another after achievement of the targets in consultation with the state governments.

16.17 Environmental Education, Training and Information

Under this broad head, grant-in-aid is provided to professional societies and NGOs for developing programmes in the areas of environmental education, wildlife and ecology. It also envisages the strengthening of the ENVIS Centres in the priority areas. A link was also provided with the Internet. The Centres of Excellence in the field of environmental education, ecological sciences, mining, environment and ornithology and natural history would be continued.

16.18 Comprehensive Legislation

Grants are released to the State Pollution Control Boards and the Department of Environment of the State Governments with the objective of strengthening their technical capabilities. Due to various decisions of the Supreme Court and the High Courts the responsibilities and commitments of the State Pollution Control Boards are increasing. A comprehensive legislation has been framed by adopting a cross medium approach; removing the multiplicity of legislations and agencies, besides removing the overlapping and ambiguous policies currently in vogue. The labelling of environmental friendly products by granting ECOMARK helps in pollution abatement. This important activity would be considerably supported during the Ninth Plan.

16.19 The National Environment Tribunal Act

The National Environment Tribunal Act, 1995 has already come into effect. It has been amended several times keeping in view the need of the hour. The principal bench of the tribunal is located at New Delhi. The supporting infrastructure for this tribunal has been divided. Recently, the Act has further been amended to make a fine tuning depending upon requirements.

16.20 Project for Environment Management Capacity Building by World Bank

In pursuance of the Environment Action Programme, the World Bank has initiated a project for environment management capacity building. The main components of the project are: environmental economics, environmental indicators, environmental law, environmental awareness and strengthening of environmental protection programmes.

16.21 Implementation of the Project

This project will be implemented through the Ministry of Environment and Forests in cooperation with the Department of Ocean Development and the Government of Gujarat. A special emphasis is being placed on Gujarat because the State is not only one of the fastest growing industrial regions of the country but is also expected to have high levels of pollution, judged by the way the industries are making a headway in the State. With the assistance of the Japanese government, studies are carried out in Surat and Delhi for the formulation of a project, specifically addressed to the concerns in the areas of air, water quality and solid waste management. An Action Plan has been prepared for regenerating the institutional structures governing urban environmental management and for identification of cost-effective technology options for improving the delivery of environmental services.

16.22 Forestry and Wildlife and Afforestation

The programmes/schemes of Ninth Five Year Plan are generally similar to those taken up during the Eighth Plan, such as Integrated Afforestation and Eco-development Project, Fuelwood and Fodder Project Scheme, Non-timber Forest Produce Scheme, Grants-in-Aid Scheme, Seed Development Scheme, etc., with greater focus and improved implementation on the basis of the experience gained.

16.22.1 Setting up a Working Group for Examining Leasing out Degraded Forests

The then Planning Commission had set up a working group to examine the prospects of leasing out of degraded forest lands to the private entrepreneurs/Forest Corporations. The main term of reference of the Working Group was to assess the economic, social and environmental feasibility of leasing or otherwise making degraded forest land available to the private entrepreneurs/Forest Corporations. The working group has submitted its report and has not recommended leasing of forest land to the private entrepreneurs either directly or indirectly through Forest Corporations.

16.22.2 Reasons for not Using Government Forests

The main reasons for not recommending the use of government forests to private industry are as follows:

- Degraded forest land leased out to industry would deprive a large populace which is dependent on these lands for their fuelwood and fodder needs.
- It will be against the interest of the farmers who wish to supply wood to industry.
- The proposal would be against the National Forest Policy 1988, the Forest Conservation Act 1980, and the Provision of Panchayats (Extension to the Scheduled Areas) Act, 1996.
- Paper and other industries consume only 10 per cent of the raw material from forests. Leasing of forest lands to these industries will adversely affect other sectors which are dependent on forests for raw materials.
- Industries will prefer to go for plantations of one or two fast growing species in place of multi-layered mixed forest which results from natural regeneration.
- Industries have shown no interest in leasing the non-forest wastelands, and therefore their plan to operate on forest lands needs careful scrutiny.

16.23 Wastelands Development

The Govt. of India envisages regeneration of wastelands to release pressures on the forests and standardization of the definition of wastelands assessment of their magnitude and their development by a reorientation of the policy of "open access" to "common property resources". Clear, quantified and phased arrangements would be evolved for an equitable sharing of the usufruct. The programmes/schemes for the Wastelands Development cover Integrated Wastelands Development Projects Scheme, Technology Development Scheme, Training and Extension Scheme and Investment Promotional Scheme, etc.

16.24 The National Forest Policy

The National Forest Policy 1988 envisages massive afforestation and social forestry programmes on all denuded, degraded and unproductive lands. Approximately, 30 m.ha of non-forest wastelands are to be brought under tree cover. This can be done by promoting farm-forestry, community forestry and government agencies, NGOs and by individuals through institutional financing. The working group on the prospects of leasing out degraded forest lands to the private entrepreneurs has recommended that:

- Private entrepreneurs may consider reclamation of non-forest wastelands which are far from habitation.
- Industries should establish direct contact with the farmers as provided in the new forest policy.
- Laws regarding ceiling of land should be liberalized to attract private entrepreneurs to take forestry projects.

16.25 Conclusion

Government of India has taken various initiatives to protect environment by introducing many schemes. These are being implemented through peoples' participation at all levels. This has created awareness among the different stakeholders who are change-agents of growth process of the economy. All our efforts are to be further strengthened so that environmental sustainability in all spheres like water, land, energy, etc., resources is maintained.

CHAPTER 17

National Institution for Transforming India (NITI) Aayog: A Glimpse

The **National Institution for Transforming India (NITI) Aayog** is a Government of India policy think-tank established by Prime Minister Narendra Modi to replace the Planning Commission.

The meaning of NITI Aayog

- NITI means policy,
- Aayog means commission.

NITI Aayog would therefore mean

- A group of people with authority entrusted by the government to formulate/regulate policies concerning transforming India.
- It is a commission to help government in social and economic issues.
- Also, it is an Institute of think tank with experts in it.

17.1 Rationale of NITI Aayog

The Planning Commission was constituted on 15.3.1950 through a Government of India Resolution, and has served India well. India, however, has changed dramatically over the past 65 years. While this has been at multiple levels and across varied scales, the biggest transformatory forces have been the following:

17.1.1 Demography

Our population has increased over three-fold to reach 125 crores. This includes an addition of over 30 crore people to Urban India. As well as an increase of 55 crore youth (below the age of 35), which is more than one and a half times the total population of the country then. Furthermore, with increasing levels of development, literacy and communication, the aspirations of our people have soared, moving from scarcity and survival to safety and surplus. We are therefore looking at a completely different India today, and our governance systems need to be transformed to keep up with the same.

17.1.2 Economy

Our economy has undergone a paradigm shift. It has expanded by over a hundred times, going from a GDP of ₹10,000 crore to ₹100 lakh crore at current prices, to emerge as one of the world's largest economics. Agriculture's share in this has seen a dramatic drop, from more than 50 per cent to less than 15 per cent of GDP. And our central government's Twelfth Five Year Plan size of ₹43 lakh crore, dwarfs the First Five Year Plan size of ₹2,400 crore. Priorities, strategies and structures dating back to the time of the birth of the Planning Commission, must thus be revisited. The very nature of our planning processes needs to be overhauled to align with this shift in sheer scale.

17.1.3 Private Enterprise

The nature of our economy, and the role of the government in it, has undergone a paradigm shift as well. Driven by an increasingly open and liberalized structure, our private sector has matured into a vibrant and dynamic force, operating not just at the international cutting edge, but also with a global scale and reach. This changed economic landscape requires a new administrative paradigm in which the role of government must evolve from simply allocating resources in a command and control eco-system, to a far more nuanced one of directing, calibrating, supporting and regulating a market eco-system. National development must be seen beyond the limited sphere of the "Public Sector". Government must thus transition from being a "provider of first and last resort" and "major player" in the economy, to being a "catalyst" nurturing an "enabling environment", where the entrepreneurial spirits of all, from small self-employed entrepreneurs to large corporations, can flourish. This importantly, frees up the government to focus its precious resources on public welfare domains such as essential entitlements of food, nutrition, health, education and livelihood of vulnerable and marginalized groups.

17.1.4 Globalization

The world at large has also evolved. Today, we live in a "global village", connected by modern transport, communications and media, and networked international markets and institutions. As India "contributes" to global dynamics, it is also influenced by happenings far removed from our borders. This continuing integration with the world needs to be incorporated into our policy making as well as functioning of government.

17.1.5 States

The States of the Union of India have evolved from being mere appendages of the Centre, to being the actual drivers of national development. The development of states must thus become the national goal, as the nation's progress lies in the progress of states. Therefore, the one-size-fits-all approach, often inherent in centralized planning, is no longer practical or efficient. States need to be heard and given the flexibility required for effective implementation. Dr B.R. Ambedkar had said with great foresight that "it is unreasonable to centralise powers where central control and uniformity is not clearly essential or is impracticable". Therefore, while emanating from global experiences and national synergy, our strategies will need to be calibrated and customized to local needs and opportunities.

17.1.6 Technology

Advancements in technology and access to information have unleashed the creative energy that emerges from the Indian kaleidoscope. They have integrated our varied regions and eco-systems in an interlinked national economy and society, opening newer avenues of coordination and cooperation. Technology is also playing a substantial role in enhancing transparency as well as efficiency, holding government more accountable. It thus needs to be made central to our systems of policy and governance.

The change and increasing mismatch has been identified for few years and many experts including the planning commission recommended some changes. They are the following:

The very first after the liberalization of 1991 – itself categorically stated that, as the role of government was reviewed and restructured, the role and functions of the Planning Commission too needed to be rethought. The Planning Commission needed to be reformed to keep up with changing trends; letting go of old practices and beliefs whose relevance had been lost, and adopting new ones based on the past experiences of India as well as other nations.

In the farewell address of the Commission in April 2014 – also urged reflection on "what the role of the Planning Commission needs to be in this new world. Are we still using tools and approaches which were designed for a different era? What additional roles should the Planning Commission play and what capacities does it need to build to ensure that it continues to be relevant to the growth process?" Mahatma Gandhi had said: "Constant development is the law of life, and a man who always tries to maintain his dogmas to appear consistent drives himself into a false position". Keeping true to this principle our institutions of governance and policy must evolve with the changing dynamics of the new India, while remaining true to the founding principles of the Constitution of India, and rooted in our *Bharatiyata* or wisdom of our civilizational history and ethos.

17.1.7 Observation of the Standing Committee on Finance

The Standing Committee on Finance of the 15th Lok Sabha observed in its 35th Report on Demand for Grants (2011–12) that the Planning Commission "has to come to grips with the emerging social realities to re-invent itself to make itself more relevant and effective for aligning the planning process with economic reforms and its consequences, particularly for the poor".

NITI Aayog is the institution to give life to these aspirations. It is being formed based on extensive consultation across the spectrum of stakeholders, including inter alia state governments, relevant institutions, domain experts and the people at large.

17.1.8 Backdrop

NITI Aayog was formed on 1st January 2015. It precedes Planning Commission. It is under the jurisdiction of Government of India. The headquarter is in New Delhi.

According to the Finance Minister of India, Arun Jaitley, "The 65-year-old Planning Commission had become a redundant organization. It was relevant in a command economy structure, but not any longer. India is a diversified country and its states are in various phases of economic development along with their own strengths and weaknesses. In this context, a "one size fits all" approach to economic planning is obsolete. It cannot make India competitive in today's global economy". Therefore, it was necessary to come up with a new institution that will formulate policies for India, and will give a boost to the Indian economy.

- 1950 – Planning commission was established.
- May 29, 2014 – The first IEO (Independent Evaluation Office) assessment report was submitted to Prime Minister Modi on May 29, three days after he was sworn in. According to Ajay Chibber, who heads the IEO, views in the report are based on the views of stakeholders and some Planning Commission members themselves. Planning Commission to be replaced by "control commission".
- August 13, 2014 – Cabinet of Modi govt. scrapped the Planning Commission
- August 15, 2014 – Mr Modi mentioned to replace Planning Commission by National Development and Reform Commission (NDRC) on the line of China.

17.2 Objectives of NITI Aayog

The salient objectives of NITI Aayog are briefly furnished below:

Cooperative and Competitive Federalism

Be the primary platform for operationalizing cooperative federalism; enabling states to have active participation in the formulation of national policy, as well as achieving time-bound implementation of quantitative and qualitative targets through the combined authority of the Prime Minister and Chief Ministers. This will be by means of systematic and structured interactions between the Union and State Governments, to better understand developmental issues, as well as forge a consensus on strategies and implementation mechanisms. The above would mark the replacement of the one-way flow of policy from centre-to-state, with a genuine and continuing Centre-State partnership. This Cooperation would be further enhanced by the vibrancy of Competitive Federalism; with the Centre competing with the states and vice versa, and the states competing, in the joint pursuit of national development.

Decentralized Planning

Restructure the planning process into a bottom-up model, empowering states, and guiding them to further empower local governments; in developing mechanisms to formulate credible plans at the village level, which are progressively aggregated up the higher levels of the government.

Shared National Agenda

Evolve a shared vision of national development priorities and strategies, with the active involvement of states. This will provide the framework "national agenda" for the Prime Minister and Chief Ministers to implement.

State's Best Friend at the Centre

Support states in addressing their own challenges, as well as building on strengths and comparative advantages. This will be through various means, such as coordinating with ministries, championing their ideas at the centre, providing consultancy support and building capacity.

Conflict Resolution

Provide a platform for mutual resolution of inter-sectoral, inter-departmental, inter-state as well as centre-state issues; facilitating consensus acceptable and beneficial to all, to bring about clarity and speed in execution.

Vision and Scenario Planning

Design medium and long-term strategic frameworks of the big picture vision of India's future – across schemes, sectors, regions and time; factoring in all possible alternative assumptions and counterfactuals. These would be the drivers of the national reforms agenda, especially focussed on identifying critical gaps and harnessing untapped potentialities. The same would need to be intrinsically dynamic with their progress and efficacy constantly monitored for necessary mid-course recalibration; and the overall environment (domestic and global) continuously scanned for incorporating evolving trends and addressing emerging challenges.

Domain Strategies

Build a repository of specialized domain expertise, both sectoral and cross-sectoral; to assist Ministries of the Central and State governments in their respective development planning as well problem-solving needs. This will especially enable the imbibing of good governance best practices, both national as well as international; especially with regards to structural reform.

Network of Expertise

Mainstream external ideas and expertise into government policies and programmes through a collaborative community of national and international experts, practitioners and other partners. This would entail being government's link to the outside world, roping in academia (universities, think tanks and research institutions), private sector expertise, and the people at large, for close involvement in the policy making process.

Sounding Board

Be an in-house sounding board whetting and refining government positions, through objective criticisms and comprehensive counter-views.

Knowledge and Innovation Hub

Be an accumulator as well as disseminator of research and best practices on good governance, through a state-of-the-art Resource Centre which identifies, analyses, shares and facilitates replication of the same.

Harmonization

Facilitate harmonization of actions across different layers of government, especially when involving cross-cutting and overlapping issues across multiple sectors; through communication, coordination, collaboration and convergence amongst all stakeholders. The emphasis will be on bringing all together on an integrated and holistic approach to development.

Internal Consultancy

Offer an internal consultancy function to central and state governments on policy and programme design; providing frameworks adhering to basic design principles such as decentralization, flexibility and a focus on results. This would include specialized skills such as structuring and executing Public Private Partnerships.

Capacity Building

Enable capacity building and technology up-gradation across government, benchmarking with latest global trends and providing managerial and technical know how.

Coordinating Interface with the World

Be the nodal point for strategically harnessing global expertise and resources coming in from across nations, multi-lateral institutions and other international organizations, in India's developmental process.

Monitoring and Evaluation

Monitor the implementation of policies and programmes, and evaluate their impact; through rigorous tracking of performance metrics and comprehensive programme evaluations. This will not only help identify weaknesses and bottlenecks for necessary course correction, but also enable data-driven policy making; encouraging greater efficiency as well as effectiveness.

17.3 Aim

The main aims of the NITI Aayog are depicted below:
- Leveraging of India's demographic dividend, and realization of the potential of youth, men and women, through education, skill development, elimination of gender bias, and employment.
- Elimination of poverty, and the chance for every Indian to live a life of dignity and self-respect.

- Redressal of inequalities based on gender bias, caste and economic disparities.
- Integrate villages institutionally into the development process.
- Policy support to more than 50 million small businesses, which are a major source of employment creation.
- Safeguarding of our environmental and ecological assets.

17.4 Terms of Reference of NITI Aayog

Monitoring and evaluation are the key mandates of the recently created National Institution for Transforming India (NITI Aayog), and official sources in the Aayog told that the government will soon move a Cabinet note proposing sweeping changes in the structure and function of the PEO to suit the current requirements of evaluation, much in line with international standards. The main changes could include the way evaluations are done besides getting domain experts to evaluate these programmes.

According to the official, the idea is to come up with a centralized terms of reference for all evaluations to be done in future besides roping in domain experts to oversee the evaluation process. "Central terms of reference will set up a benchmark for evaluating officials to grade ministries and outcomes on set parameters. Evaluation will be an ongoing process and not a one-off exercise done on one programme at a given time. Over a period of time, this will ensure that the Centre even starts monitoring the effectiveness with which ministries and agencies implement the programme," the official added. The Cabinet resolution of January 1 that led to the setting up of NITI Aayog has clearly outlined monitoring and evaluation as key objectives of the Aayog.

"The NITI Aayog will work towards actively monitoring and evaluating the implementation of programmes and initiatives, including the identification of the needed resources so as to strengthen the probability of success and scope of delivery," it had said.

17.5 Composition of NITI Aayog

Composition of NITI Aayog and its members are: (i) Prime Minister of India as the Chairperson, (ii) Governing Council comprising the Chief Ministers of all the States and union territories with legislatures and lieutenant governors of other Union Territories. (iii) Regional Councils will be formed to address specific issues and contingencies impacting more than one state or a region. These will be formed for a specified tenure. The Regional Councils will be convened by the Prime Minister and will be composed of the Chief Ministers of States and Lt. Governors of Union Territories in the region. These will be chaired by the Chairperson of the NITI Aayog or his nominee, (iv) Experts, specialists and practitioners with relevant domain knowledge as special invitees nominated by the Prime Minister, (v) Full-time organizational framework (in addition to Prime Minister as the Chairperson) comprising.

(a) Vice-Chairperson: Arvind Panagariya (resigned)
(b) Members: Two full-time: economist Bibek Debroy and former DRDO chief V.K. Saraswat

(c) Part-time members: Maximum of two from leading universities research organizations and other relevant institutions in an ex-officio capacity. Part-time members will be on a rotational basis

(d) Ex Officio members: Maximum of four members of the Union Council of Ministers to be nominated by the Prime Minister

(e) Chief Executive Officer: To be appointed by the Prime Minister for a fixed tenure, in the rank of Secretary to the Government of India. Sindhushree Khullar appointed as the Chief Executive Officer

(f) Secretariat as deemed necessary

17.6 Specialized Wings of NITI Aayog

NITI Aayog will also house many specialized wings. These are furnished below in brief:

- **Research Wing** that will develop in-house sectoral expertise as a dedicated think tank of top notch domain experts, specialists and scholars.
- **Consultancy Wing** that will provide a market-place of whetted panels of expertise and funding, for Central and State Governments to tap into; matching their requirements with solution providers, public and private, national and international. By playing match-maker instead of providing the entire service itself, NITI Aayog will be able to focus its resources on priority matters, providing guidance and an overall quality check to the rest.
- **Team India Wing** comprising representatives from every state and ministry, will serve as a permanent platform for national collaboration. Each representative will:
  (a) Ensure every state/ministry has a continuous voice and stake in the NITI Aayog.
  (b) Establish a direct communication channel between the state/ministry and NITI Aayog for all development related matters, as the dedicated liaison interface.

A national Hub-Spoke institutional model will be developed, with each state and ministry encouraged to build dedicated mirror institutions, serving as the interface of interaction. These institutions, in turn, will nurture their own networks of expertise at state and ministry level.

17.7 Present Members

- Chairperson: Prime Minister Narendra Modi
- CEO: Amitabh Kant
- Vice Chairperson: Rajiv Kumar
- Ex-Officio Members: Rajnath Singh, Arun Jaitley, Suresh Prabhu and Radha Mohan Singh
- Special Invitees: Nitin Gadkari, Smriti Zubin Irani and Thawar Chand Gehlot
- Full-time Members: Bibek Debroy, V.K. Saraswat and Ramesh Chand
- Governing Council: All Chief Ministers and Lieutenant Governors of Union Territories

17.8   Difference Between NITI Aayog and Planning Commission

Sl.	NITI Aayog	Planning Commission
Financial Clout		
1.	(a) To be an advisory body, or a think-tank. (b) It does not hold power to allocate funds. (c) The powers to allocate funds might be vested in the finance ministry.	Enjoyed the powers to allocate funds to ministries and state governments.
Full-time members		
2.	The number of full-time members could be fewer than Planning Commission.	The last Commission had eight full-time members in it State's role.
Member secretary		
3.	To be known as the CEO and to be appointed by the prime minister.	Secretaries or member secretaries were appointed through the usual process.
Part-time members		
4.	To have many part-time members, depending on the need from time to time.	Full Planning Commission had no provision for part-time members.
Organization		
5.	(a) New posts of CEO, of secretary rank, and Vice-Chairperson. (b) It will also have five full-time members and two part-time members. (c) Four cabinet ministers will serve as ex-officio members.	Had deputy chairperson, a member secretary and eight full-time members.
Constitution		
6.	Governing Council has state chief ministers and lieutenant governors.	The commission reported to National Development Council that had state chief ministers and lieutenant governors.
Participation		
7.	(a) Consulting states while making policy and deciding on funds allocation. (b) Final policy would be a result of that.	Policy was formed by the commission and states were then consulted about allocation of funds.
Allocation		
8.	No power to allocate funds.	Had power to decide allocation of government funds for various programmes at national and state levels.
Nature		
9.	NITI is a think-tank and does not have the power to impose policies.	Imposed policies on states and tied allocation of funds with projects it approved.

17.9 State Government V/S Central Government

Planning Commission was envisaged as a mere departmental body of the Central government, where there was no representation from the states. This did not matter in the beginning when the same political Party ruled at the Centre and in the states; but it later became a serious limitation of the planning process, since this process came into conflict with the federal nature of the Indian polity. The Manmohan Singh government had already initiated the process by adopting the bizarre stratagem of inviting a neo-liberal economist who had been a Fund-Bank employee to head the Commission; the Modi government has gone the whole length and has completed this process. In fact, the winding up of the Planning Commission itself marks a major step towards the consolidation of a neo-liberal State.

There are at least two ways in which centralization of economic power will increase under the new dispensation.

(i)   First, the winding up of the Planning Commission will inevitably mean a strengthening of the Ministry of Finance, which is a far more closely controlled Departmental body of the Central government than the Planning Commission of ever was.

(ii)   There were typically three channels for the devolution of resources from the centre to the states in India:

(a)   through the Finance Commission which, though a constitutional body, was always appointed by the central government, with no consultations with the states, and hence filled with persons willing to do its bidding;

(b)   through the Planning Commission which again was a departmental body, though admittedly of an unconventional kind, of the central government; and the

(c)   through the Ministry of Finance which was a conventional departmental body and which made financial transfers to states at its own discretion.

(iii)   While the centre influenced all three channels of transfers, these three channels can clearly be ordered in terms of their relative independence from the dictates of the central government, the last of these being obviously the one that is directly governed by such dictates.

(iv)   The winding up of the Planning Commission will necessarily mean therefore that the flows which used to come to the states through the Planning Commission channel will now be effected through the Ministry of Finance; and this will mean greater direct control by the Centre over what flows to which state.

(v)   The second reason that the winding up of the Planning Commission, even in the form it existed under the Manmohan Singh government, will lead to centralization is the simultaneous abolition of an apex body, the National Development Council.

The National Development Council, to which the Planning Commission reported, though not a statutory body, was a forum where state Chief Ministers expressed themselves, not just on issues affecting their own states but on national development issues. True, the

NDC did not vote; but the Centre was under some pressure at its meetings to accommodate states' demands (though no doubt there certain notable instances where it did not). What is more, the states came to learn of each other's positions at the NDC meetings and often derived confidence from the fact that other states too were voicing concerns like their own. But now, according to the information made available so far at any rate, there will be no NDC, but only a few regional councils where the prime minister will sit with the state chief ministers.

A bunch of supplicant state governments of regions will be pleading for greater largesse from the Centre at occasional regional meets. Since neo-liberalism presents itself in a "State versus Market" context, as a rolling back of State intervention, and its replacement by greater reliance on "the market", it creates the impression that it entails more scattered, more decentralized, more dispersed economic power. Indeed, the term 'liberal' in "neo-liberal" (which alas one must use only because of its current prevalence) reinforces this impression.

But this impression is completely wrong. Neo-liberal economic policies to do not with the "State versus market" dichotomy but with a change the State whereby it seeks almost exclusively to promote the interests of the corporate-financial oligarchy at the expense of the vast mass of urban workers, agricultural labourers, peasants and petty producers. Neo-liberalism refers not to the area of intervention by the State, but to the class nature of that intervention (from which of course the area of intervention is derived). If the State seeks to privatize nationalized banks, then the reason is not because it believes in the "market" in some abstract sense, but because it wants to hand over control over banks to the corporate-financial oligarchy allied to international finance capital. Such a shift the State necessarily requires centralization of political power; it also requires centralization of economic power within any federal polity (if the federal polity itself exists and is not destroyed as in Yugoslavia) in favour of the federal authority at the expense of the states. The fact that "Economic liberalism" is necessarily associated with political authoritarianism, i.e. with an attenuation in a myriad way of democratic institutions and democratic rights of the people (of which the "ordinance raj" of today is a classic example), has been widely recognized, including even by the renowned conservative American economist, the late Paul Samuelson. But the centralization of economic power with the federal authority at the expense of the states within any existing federal polity (except when the breaking up of the country is even more "attractive" to international finance capital), is less discussed and recognized. But a moment's consideration should clarify why such centralization is required.

This is precisely what the Modi government is attempting to do in its bid to carry forward "reforms" on behalf of the corporate-financial oligarchy. The substitution of the Planning Commission by the NITI Aayog is not just a means of providing greater elbow room to the corporate-financial oligarchy; it is simultaneously a means of curbing the states' economic powers. The neo-liberal State whose consolidation it carries forward is simultaneously a highly centralized State in terms of political and economic authority.

17.10 Significant Achievements of NITI Aayog Over the Last Three Years

(i) **Vision Document, Strategy and Action Agenda beyond 12th Five Year Plan:** Replacing the Five-Year Plans beyond 31st March 2017, NITI Aayog is in the process of preparing the 15-year vision document keeping in view the social goals set and/or proposed for a period of 15 years; A 7-year strategy document spanning 2017–18 to 2023–24 to convert the longer-term vision into implementable policy and action as a part of a "National Development Agenda" is also being worked upon. The 3-year Action Agenda for 2017–18 to 2019–20, aligned to the predictability of financial resources during the 14th Finance Commission Award period, has been completed and will be submitted before the Prime Minister on April 23rd at the 3rd Governing Council Meeting.

(ii) **Reforms in Agriculture**

(a) **Model Land Leasing Law**

Taking note of increasing incidents of leasing in and out of land and suboptimal use of land with lesser number of cultivators, NITI Aayog has formulated a Model Agricultural Land Leasing Act, 2016 to both recognize the rights of the tenant and safeguard interest of landowners. A dedicated cell for land reforms was also set up in NITI. Based on the model act, Madhya Pradesh has enacted separate land leasing law and Uttar Pradesh and Uttarakhand have modified their land leasing laws. Some States, including Odisha, Andhra Pradesh and Telangana, are already at an advance stage of formulating legislations to enact their land leasing laws for agriculture.

(b) **Reforms of the Agricultural Produce Marketing Committee Act**

NITI Aayog consulted with the States on 21 October 2016 on three critical reforms like (i) Agricultural marketing reforms, (ii) Felling and transit laws for tree produce grown at private land and (iii) Agricultural land leasing. Subsequently, Model APMC Act version 2 prepared. States are being consulted to adopt APMC Act version 2.

(c) **Agricultural Marketing and Farmer-Friendly Reforms Index**

NITI Aayog has developed the first ever "Agriculture Marketing and Farmer-Friendly Reforms Index" to sensitise states about the need to undertake reforms in the three key areas of Agriculture Market Reforms, Land Lease Reforms and Forestry on Private Land (Felling and Transit of Trees). The index carries a score with a minimum value "0" implying no reforms and maximum value "100" implying complete reforms in the selected areas.

As per NITI Aayog's index, Maharashtra ranks highest in implementation of various agricultural reforms. The State has implemented most of the marketing reforms and offers the best environment for undertaking agri-business among all the States and UTs. Gujarat ranks second with a score of 71.50 out of 100, closely followed by Rajasthan and Madhya Pradesh. Almost two third States have not been able to reach even the halfway mark of reforms score, in the year 2016–17. The index aims to induce a healthy competition between States and percolate best practices in implementing farmer-friendly reforms.

(iii) **Reforming Medical Education**

A committee chaired by Vice Chairman, NITI Aayog recommended scrapping of the Medical Council of India and suggested a new body for regulating medical education. The draft legislation for the proposed National Medical Commission has been submitted to the Government for further necessary action.

(iv) **Digital Payments Movement**

(a) An action plan on advocacy, awareness and co-ordination of handholding efforts among public, micro enterprises and other stakeholders was prepared. Appropriate literature in print and multimedia was prepared on the subject for widespread dissemination. Presentations/interactions were organized by NITI Aayog for training and capacity building of various Ministries/Departments of Government of India, representatives of State/UTs, Trade and Industry Bodies as well as all other stakeholders.

(b) NITI Aayog also constituted a Committee of Chief Ministers on Digital Payments on 30th November 2016 with the Chief Minister of Andhra Pradesh, Chandrababu Naidu, as the Convener to promote transparency, financial inclusion and a healthy financial ecosystem nationwide. The Committee submitted its interim report to Hon'ble Prime Minister in January 2017.

(c) To incentivize the States/UTs for promotion of digital transactions, Central assistance of ₹50 crore would be provided to the districts for undertaking information, education and communication activities to bring 5 crore Jan Dhan accounts to digital platform.

(d) Cashback and referral bonus schemes were launched by the Prime Minister on 14.4.2017 to promote the use of digital payments through the BHIM App.

(e) Niti Aayog also launched two incentive schemes to promote digital payments across all sections of society – the Lucky Grahak Yojana and the Digi Dhan Vyapar Yojana – Over 16 lakh consumers and merchants have won ₹256 crore under these two schemes.

(f) Digi Dhan Melas were also held for 100 days in 100 cities, from December 25th to April 14th.

(v) **Atal Innovation Mission**

The Government has set up Atal Innovation Mission (AIM) in NITI Aayog with a view to strengthen the country's innovation and entrepreneurship ecosystem by creating institutions and programmes that spur innovation in schools, colleges, and entrepreneurs in general. In 2016–17, the following major schemes were rolled out:

(a) Atal Tinkering Labs (ATLs): To foster creativity and scientific temper in students, AIM is helping to establish 500 ATLs in schools across India, where students can design and make small prototypes to solve challenges they see around them, using rapid prototyping technologies that have emerged in recent years.

(b) Atal Incubation Centres (AICs): AIM will provide financial support of ₹10 crore and capacity building for setting AICs across India, which will help start-ups expand quicker and enable innovation-entrepreneurship, in core sectors such as manufacturing, transport, energy, education, agriculture, water and sanitation, etc.

(vi) **Indices Measuring States' Performance in Health, Education and Water Management:** As part of the Prime Minister's Focus on outcomes, NITI has come out with indices to measure incremental annual outcomes in critical social sectors like health, education and water with a view to nudge the states into competing for better outcomes, while at the same time sharing best practices and innovations to help each other – an example of competitive and cooperative federalism.

(vii) **Sub-Group of Chief Ministers on Rationalization of Centrally Sponsored Schemes:** Based on the recommendations of this Sub-Group, a Cabinet note was prepared by NITI Aayog which was approved by the Cabinet on 3rd August 2016. Among several key decisions, the sub-group led to the rationalization of the existing CSSs into 28 umbrella schemes.

(viii) **Sub-Group of Chief Ministers on Swachh Bharat Abhiyan**

Constituted by NITI Aayog on 9th March 2015, the Sub-Group has submitted its report to the Hon'ble Prime Minister in October 2015 and most of its recommendations have been accepted.

(ix) **Sub-Group of Chief Ministers on Skill Development**

Constituted on 9th March 2015, the report of the Sub-Group of Chief Ministers on Skill Development was presented before the Hon'ble Prime Minister on 31st Dec. 2015. The recommendation and actionable points emerging from the report were approved by the Hon'ble Prime Minister and are in implementation by the Ministry of Skill Development.

(x) **Task Force on Elimination of Poverty in India**

Constituted on 16th March 2015 under the Chairmanship of Dr Arvind Panagariya, Vice Chairman, NITI Aayog, the report of the Task Force was finalized and submitted to the Prime Minister on 11th July 2016. The report of the Task Force primarily focusses on issues of measurement of poverty and strategies to combat poverty. Regarding estimation of poverty, the report of the Task Force states that "a consensus in favour of either the Tendulkar or a higher poverty line did not emerge. Therefore, the Task Force has concluded that the matter be considered in greater depth by the country's top experts on poverty before a final decision is made. Accordingly, it is recommended that an expert committee be set up to arrive at an informed decision on the level at which the poverty line should be set." With respect to strategies to combat poverty, the Task Force has made recommendations on faster poverty reduction through employment intensive sustained rapid growth and effective implementation of anti-poverty programmes.

(xi) **Task Force on Agriculture Development**

The Task Force on Agricultural development was constituted on 16th March 2015 under the Chairmanship of Dr Arvind Panagariya, Vice Chairman, NITI Aayog. The Task Force based on its works prepared an occasional paper entitled "Raising Agricultural Productivity and Making Farming Remunerative for Farmers" focussing on 5 critical areas of Indian Agriculture. These are (i) Raising Productivity, (ii) Remunerative Prices to Farmers, (iii) Land Leasing, Land Records and Land Titles, (iv) Second Green Revolution-Focus on Eastern States, and (v) Responding to Farmers' Distress. After taking inputs of all the States on occasional paper and through their reports, the Task Force submitted the final report to Prime Minister on 31st May 2016. It has suggested important policy measures to bring in reforms in agriculture for the welfare of the farmers as well as enhancing their income.

(xii) **Transforming India Lecture Series**

As the government's premier think-tank, NITI Aayog views knowledge building and transfer as the enabler of real transformation in States. To build knowledge systems for States and the Centre, NITI Aayog launched the "NITI Lectures: Transforming India" series, with full support of the Prime Minister on 26th August 2016. The lecture series is aimed at addressing the top policy making team of the Government of India, including members of the cabinet and several top layers of the bureaucracy. It aims is to bring cutting edge ideas in development policy to Indian policy makers and public, to promote the cause of transformation of India into a prosperous modern economy. The Deputy Prime Minister of Singapore, Shri Tharman Shanmugaratnam, delivered the first lecture on the topic: "India and the Global Economy". On 16th November 2016, Bill Gates, Co-Founder, Bill and Melinda Gates Foundation, delivered the second lecture in the series under the theme: "Technology and Transformation".

17.11 Criticism

The government's move to replace the Planning Commission with a new institution called "NITI Aayog" was criticized by opposition parties of India. The Congress sought to know whether the reform introduced by the BJP-led government was premised on any meaningful programme or if the move was simply born out of political opposition to the party that ran the Planning Commission for over 60 years. Most of the country see this as continuation of the negativism and policy paralysis approach of the Congress. "The real issue is do you (the government) have a substantive meaningful programme to reform the Planning Commission?" Congress spokesperson Abhishek Manu Singhvi said. "If you (the BJP government) simply want to abolish it (the commission), because it is something which (Jawaharlal) Nehru created for this country and you don't like Nehru or simply because it was run by the Congress for 60 years and you don't like the Congress, that is pitiable," he said.

The Communist Party of India-Marxist said a mere change in the name would not yield the desired results. "Mere changing this nomenclature, is not going to serve the purpose. Let us wait and see what the government is eventually planning".

CPI-M leader Sitaram Yechury said. These statements have been branded by several as motivated because the Communist Party of India-Marxist and their red brethren have a history of anti-nationalism, having collaborated with the British during the Quit India movement and supported China during the Indo-China war. Under the NDA and PM Modi's decisive leadership, India is set to overtake the growth rate of China by 2016 and this may not be palatable to the Red comrades in India who might be secretly rooting for China's continuing dominance.

"The Planning Commission used to plan policy. I don't know what the government is trying to do by merely changing the nomenclature from Planning Commission to Niti Aayog," said Congress spokesman Manish Tewari. However, former Commerce and Industry Minister Nirmala Sitharaman of BJP accused the critics of being "ignorant of facts".

"With the new set of changes, the state governments no longer need to have a begging attitude and instead take independent steps for development," said Sitharaman. Through this the NDA government is fulfilling one more of its key promises of robust federalism.

"The idea to create an institution where states' leaders will be part and parcel of the collective thinking with the Centre and other stakeholders in formulating a vision for the development of the country is right on as compared with the previous structure, where a handful of people formulated the vision and then presented it to the National Development Council (NDC). This was not entirely absorbed and adopted by the latter," said former Planning Commission member Arun Maira.

17.12 Conclusion

Most of the people think that it is a political decision, to replace Planning Commission by NITI Aayog. But looking at functions and objectives of the NITI Aayog, we can say that it is better than Planning Commission. It will be useful for different states, as it gives power to allocate funds to the states themselves. Aayog is going to frame an integrated energy policy soon, which will cover all related aspects including climate change issues. Therefore, the NITI Aayog would be able to accelerate the economic growth of the country. It would also develop all parts of the country.

References

The NITI Aayog by Prabhat Patnaik.
Tactful Management Research Journal.
Press Information Bureau.
From Planning to NITI Transforming (Govt. of India).

Bibliography

Karmakar, K.G. and Banerjee, Gangadhar. *Rural Development: Concepts and Strategies for India*. New Delhi: Northern Book Centre, 2015.

Banerjee, G.D. and Banerji, Srijeet. *Skills and Entrepreneurship Development*, New Delhi, Ane Books, 2015.

—————————. *The Economics of Financial Inclusion*. New Delhi: Ane Books Pvt. Ltd., 2016.

—————————. *Fresh, Brackish and Coastal Water Fish Sector: Challenges and Opportunities*. New Delhi: Ane Books, 2017. Under print.

—————————. *The Economics of Sustainable Agriculture and Alternative Production System*. New Delhi: Ane Books, 2016.

Karmakar, K.G. and Bandopadhyay, G.D. *Commercial Agriculture in India: Post Harvest Needs of Small Holder Farmers*. USA: Horizon research publishing, 2017.

Banerjee, G.D. and Banerji, Srijeet. *Mango Cultivation and Marketing in India*. New Delhi: Abhijeet Publications, 2012.

Banerjee, G.D. *Rural Entrepreneurship Development Programme: An Impact Assessment*. New Delhi: Abhijeet Publications, 2012.

Banerjee, G.D. and Banerji, Srijeet. *Perspective on Indian Agricultural Development*. New Delhi: Abhijeet Publications, 2011.

—————————. *Issues on Rural Finance, Infrastructure and Rural Development*. New Delhi: Abhijeet Publications, 2010.

—————————. *Nagaland: The Land of Immense Potential*. New Delhi: Abhijeet Publications, 2010.

Banerjee, G.D. *Tea Industry in Transition*. New Delhi: Abhijeet Publications, 2009.

Banerjee, G.D. and Banerji, Srijeet. *Tea Plantation Industry: A Road Map Ahead*. New Delhi: Abhijeet Publications, 2009. Print.

—————————. *Tea Trade: Dimensions and Dynamics*. New Delhi: Abhijeet Publications, 2008. Print.

Banerjee, G.D. *Export Potential of Indian Tea*. New Delhi: Abhijeet Publications, 2008.

—————————. *Rural Entrepreneurship Development Programme: An Impact Assessment*. Occasional Paper 57. Department of Economic Analysis and Research, National Bank for Agriculture and Rural Development, Mumbai, 2011.

Banerjee, G.D. and Palke, L.M. *Economics of Pulses Production and Processing of India.* Occasional Paper 51. Department of Economic Analysis and Research, Mumbai, 2010.

Karmakar K.G. and Banerjee, G.D. *The Tea Industry in India: A Survey.* Occasional Paper 39. Department of Economic Analysis and Research, National Bank for Agriculture and Rural Development, Mumbai, 2005.

Karmakar, K.G., Banerjee, G.D. and N.P. Mohapatra. *Towards Financial Inclusion in India.* New Delhi: SAGE Publications, 2011.

Banerjee, G.D. and Banerjee, Sarda. *Sustainable Tea Plantation Management.* Lucknow: International Book Distributing Company, 2008.

Banerjee, G.D. and Banerji, Srijeet. *Darjeeling Tea: The Golden Brew.* Lucknow: International Book Distributing Company, 2007.

Hundekar, S.G., Mohan Das, V.B. and Banerjee, G.D. (Eds.) *Challenges Before Small Scale Industries.* Allahabad: Horizon Publishers, 1997.

Banerjee, Gangadhar. *Tea Plantation Industry in India between 1850 and 1992: Structural Changes.* Assam: Lawyers Book Stall, 1996. Print.

Reports

Banerjee Gangadhar, "Socio-Economic Implications of high Yielding Varieties of Food Grains in the Eastern Region of India." UNDP Global Project, Agro Economic Research Center. Santiniketan, Visva Bharati, 1974.

Banerjee Gangadhar, "Efficiency of Small Farms under Different Tenurial Conditions with Special reference to North Bengal," ICAR, New Delhi, Dept. of Economics and Politics, Visva Bharati, Santiniketan, 1976.

Tea Board, Techno-Economic Survey of Darjeeling Tea Industry. Tea Board of India, Kolkata, 2001.

Tea Board, Techno-Economic Survey of Tripura Tea Industry. Tea Board of India, Kolkata, 1978.

Tea Board, "Techno-Economic Survey of Small Tea Gardens in Kottayam and Idukki" (Kerala). Tea Board of India, Kolkata, 1979.

Tea Board, "Techno-Economic Survey of Small Tea Gardens in Kangra (H.P.). Tea Board of India," Kolkata, 1979.

Tea Board, "Techno-Economic Survey of Nilgiris Tea Industry," Tea Board of India, Kolkata, 1980.

Tea Board, "Techno-Economic Survey of Cachar Tea Industry," Tea Board of India, Kolkata, 1981.

National Bank for Agriculture and Rural Development, Evaluation study report on Inland Fisheries in Nadia District, Kolkata, 1987.

National Bank for Agriculture and Rural Development, Evaluation study report on Inland Fishery in West Tripura District, Kolkata, 1992.

National Bank for Agriculture and Rural Development, Evaluation study report on Poultry Farming in (Broiler) in Midnapur District, Kolkata, 1999.

Papers in International Conferences/Seminars

"Innovative Business & Technology Strategies For Developing Countries." Conducted by Asian Management Science Association & Putra Intelek International College, Malaysia at Dubai, December 2011. Published in Conference volume.

"Broad Based Growth through Rural Development: A Study of Rural Entrepreneurship Development Programme of NABARD and its Viability." Presented and published in the International Conference of Facets of Business Excellence. Organized by Essex University and Institute of Management Technology (IMT), Ghaziabad, November 2011.

"Financial Inclusion: Challenges and Opportunities." Presented in the International Conference of Financial Inclusions – Multi purposive. Organized by Institute for Technology and Management (ITM), Mumbai in association with NABARD, Mumbai on 3 and 4 February, 2012.

"Self-help Group – A Peoples' Institution providing Economic and Social Entitlement with the Poor," Vol. 6, Issue 1, July 2012. Published in Sinhgad International Business Review, International Research Journal of Sinhgad Institute of Management, Pune, January 2013.

"Entrepreneurship Development Programme – A Model of Financial Inclusion." Presented in the International Conference on "Inclusive Growth Through Innovative HR Practices and Alternative Finance, 20 February 2015. Published in Conference Issue. Organized by Institute for Technology and Management (ITM), Mumbai.

Papers in National Conferences/Seminars

"Role of Information and Communication Technology in Dissemination of Knowledge in Agriculture – Its Efficacy and Scope." Presented and published in 47th Annual Conference of Indian Journal of Agricultural Economics. Organized by Indian Society of Agricultural Economist in collaboration of University of Agricultural Sciences, Dharwad, Indian Council of Agriculture Research, Ministry of Agriculture, Govt. of India, November 2011.

"Business Continuity Plan – A Critical Review." Presented in National Seminar on "Business Innovative Strategies: Breakthrough while Break down." Organized by Lala Lajpat Rai Institute of Management Studies and Research, Mumbai, September 2011.

"Role of Small and Medium Enterprises in Accelerating Economic Development of India." Presented in the National Seminar on "Economic Environment and Business Sustenance." Organized by Vivekanand Education Society's Institute of Management Studies and Research, Mumbai, 23 and 24 February, 2012.

"Crop Diversification – An Analysis" and "Product Diversification of Tea – A Value Addition." Presented in the Regional Seminar on Diversification of Agriculture. Organized by Agro-economic Research Centre, Visva Bharati, Santiniketan in collaboration with Indian Journal of Agricultural Society, 23–25 March 2012 at Santiniketan, Birbhum, West Bengal and published in Conference volume.

Papers published in the following Journals

"Economics of Poultry Farming in Andhra Pradesh: A Field Enquiry. Annual Conference. *Indian Journal of Agricultural Economics*, Vol. 69, No. 3, July-September (2014).

"Role of Information and Communication Technology in Dissemination of Knowledge in Agriculture Sector: Its Efficiency and Scope." Annual Conference. *Indian Journal of Agricultural Economics*, July-September (2011).

"Impact of Agro-processing on Incomes of Primary Producers." Annual Conference. *Indian Journal of Agricultural Economics*, Dec. (2003).

"Changes Overtimes in the Levels of Income of the Weaker Section of Rural Community (WB)." Annual Conference. *Indian Journal of Agricultural Economics*, Issue June-July (1972).

"Economics of Mango Orchards and Processing: A Field Enquiry." *Supply Chain Pulse* Vol. 2, No. 3, July-September (2011).

"Supply Chain Management: A Critical Review." *Supply Chain Pulse* Vol. 6, No. 3, October-December (2015).

"Rural Entrepreneurship Development Programme: An Impact Assessment." *Journal of Bangiya Arthaniti Parishad*, No. 18, March (2010).

"Global Meltdown: A Wake up Call for the Economy." *Journal of Research* Vol. 4, No. 2, July-December (2012).

"Economics of Mango Processing: A Field Investigation, *Journal of Research* Vol. 3, July-December (2010).

"Coal Ash in India: Opportunities and Challenges." *Journal of Research*, Vol. 2, January-June (2010).

"Commodity Derivatives Markets: Opportunities and Challenges." *Journal of Research*, July-December (2009).

"Individual Investment Pattern – A Dynamic Process, Regular Monitoring, and Re-valuation – An Empirical Investigation." *Journal of Development Research,"* Vol. 3, No. 1, April-June (2011).

"Indian Planning – Past and Present." *Journal of Development Research",* Vol. 3, No. 2, July-September (2011).

"Micro-finance in India – A Road Map Ahead." *Journal of Development Research,"* Vol. 3, No. 4, January-March (2012).

"Role of Different Channels, Market Arrivals and Price Spread: A Case for Mango." *Journal of Development Research,* Vol. 3, No. 2, January-March (2012).

"Sustainable Agriculture: Issues and Strategies." *Journal of Development Research,"* Vol. 4, No. 1, April-June (2012).

"Product Diversification and Value Addition of Tea." *Indian Association of Social Science Institutions (IASSI) Quarterly Journal,* October-December (2006).

"China Pure Tea: Tea with Miraculous Functions." *Indian Association of Social Science Institutions" (IASSI) Quarterly Journal,* Vol. 24, No. 2, October-December (2005).

"Sustainable Tea Plantation Management." *Indian Association of Social Science Institutions" (IASSI) Quarterly Journal,* Vol. 22, No. 3, Jan-March (2004).

"Agro-processing: A Link between Agriculture and Industry." *Indian Association of Social Science Institutions (IASSI) Quarterly Journal,* Vol. 22, No. 3, Jan-March (2004).

"Indian Tea: A No Longer A Fashionable Slogan." *Indian Association of Social Science Institutions" (IASSI) Quarterly Journal,* Vol. 19, No. 4, April-June (2002).

"Agricultural Extension: A Critical Analysis of Various Models." *Man, and Development,* Vol. 33, No. 3, September (2001).

"Horticulture Strives Well in Food and Nutritional Security." *Financing Agriculture,"* Vol. 32, No. 1, April-June (2000).

"Agricultural Extension: Assessment of Various Models." *Financing Agriculture,"* Vol. 32, No. 2, October-December (2000).

"Biotechnology for Indian Farmers." *Financing Agriculture,* Vol. 33, No. 3, July-September (2001).

"Water Resources Management: A Long-Term Planning and Judicious Use of Water and Food Security." *Financing Agriculture,"* Vol. 33, No. 4, October-December (2001).

"Evaluation Studies of Self-Help Groups." *Financing Agriculture,* Vol. 34, No. 2, April-June (2002).

"Economics of Medicinal Plants." *Financing Agriculture,* Vol. 34, No. 3, July-September (2003).

"Boosting of Agricultural Exports: Opportunities and Challenges." *Financing Agriculture*, January-March (2005).

"Marketing of Indian Tea: A Review." *Indian Journal of Agricultural Marketing*, Vol. 33, No. 3, Sept (2001).

"Marketing of Mango Processing under Institutional Finance: A Case Study in Andhra Pradesh." *Afro-Asian Journal of Rural Development*, Vol. 34, No. 2, July-December (2001).

"Agro-processing Sector: Some Issues." *National Bank News Review*, July-September (2004).

"Industrialization Sets Back to Farm Income." *National Bank News Review*, January-March (2003).

"Economics of Green Tea." *National Bank News Review*, April-June (2002).

"Indian Tea in Syria." *National Bank News Review*, October-December (2002).

"Indian Tea in Jordan." *National Bank News Review*, July-September (2001).

"Indian Tea in Libya." *National Bank News Review*, July-September (1999).

"National Bank's Role for Tea Development." *National Bank News Review*, January-March (1997).

"The Role of National Bank for Development of Tea Industry." *National Bank News Review*, October-December (1995).

"Dry and Rain Fed Farming." *National Bank News Review*, October-December (1993).

"Potentials of Horticulture Crops." *National Bank News Review*, April-June (1992).

"Market Potential of Indian Tea in Syria." *National Bank News Review*, September-November (1991).

"Cashew nut Plantation." *National Bank News Review*, January-March (1990).

"A Fresh Insight into the Problems of Rural Development in the North-East Region." *National Bank News Review*, July-September (1988).

"Green Tea: Peep into the Past." 07 October-December *National Bank News Review*, April-June (1987).

"Rain-fed and Dry Land Farming: A Balanced Agricultural Strategy Vital." *National Bank News Review*, April-June (1986).

"Brackish Water Fish Culture: A Cast Study." *National Bank News Review*, January-March Issue (1985).

"Agricultural Economy of Nagaland" – Part I, Part II & Part III. *E-Journal, Agricultural and Rural Development*.

"Horticulture Over View: Poised for Golden Revolution." *E-Journal, Agricultural and Rural Development,* November-December Issue (2002).

"Financing Horticulture." *E-Journal, Agricultural and Rural Development,* January-February Issue (2003).

"A Difficult Terrain." *E-Journal, Agricultural and Rural Development,* July-August Issue (2003).

"Water Vision." *Wastelands News: A Quarterly Newsletter of Society for Promotion of Wastelands Development.* November 2004-January 2005.

"Corporate Partnership for Agricultural Development." *Kurukshetra,* Vol. 53, No. 6, April (2005).

"Eco -Labelling: A Major Thrust." *Kurukshetra,* January (2006).

"Tea Industry at the Crossroads."*Yojana,* Vol. 49, 15 August 2005.

"Economics of Organic Farming." *Agriculture Today,* Vol. 8, No. 10, October (2005).

"Eco-Labelling in Agriculture." *Agriculture Today,* Vol. 8, No. 12, December (2005).

"Problems and Prospects of Tarai Tea Industry." *TAI Newsletter,* Vol. 16, No. 2, February (1983).

"Problems and Prospects of Indian Tea Exports." *TAI Newsletter,* Vol. 16, No. 3, March (1983).

" The Role of FAO/UNCTAD and ITPA for Development of World Tea." *TAI Newsletter,* Vol. 16, No. 5, May (1983).

" Export Prospects of Indian Tea in Tunisia." *TAI Newsletter,* Vol. 16, No. 7, July (1983).

"Tea Plantation of India (North)." *TAI Newsletter,* Vol. 16, No. 9/10, September-October (1983).

"Structural Development in Tea Industry." *TAI Newsletter,* Vol. 16, No. 11, November (1983).

"Evolution of Tea Industry in India" (Part I). *TAI Newsletter,* Vol. 16, No. 12, December (1983).

"Evolution of Tea Industry in India" (Part II). *TAI Newsletter,* Vol. 17, No. 2, February (1984).

"Ownership of Tea Estates." *TAI Newsletter,* Vol. 17, No. 3, March (1984).

"Japan As A Base of Indian Tea." *TAI Newsletter,* Vol. 17, No. 3, August (1984).

"Changes in Pattern of Management." *TAI Newsletter,* Vol. 8, No. 9, September (1984).

"Area, Production and Yield of Tea in Different Excise Zones." *TAI Newsletter,* Vol. 17, No. 12, December (1984).

"Changes in the Pattern of Ownership." *TAI Newsletter*, Vol. 19, No. 2, February (1985).

"Market Research." *TAI Newsletter*, Vol. 20, No. 3, March (1985).

"The Role of Tea in Indian Economy." *TAI Newsletter*, Vol. 21, No. 4, April (1985).

"Export Prospects of Indian Tea in Nigeria." *TAI Newsletter*, Vol. 21, No. 6, June (1985).

"Export Prospects of Indian Tea in Singapore." *TAI Newsletter*, Vol. 21, No. 11, November (1985).

"Tea Plantation in South India." *TAI Newsletter*, Vol. 22, No. 10, October (1986).

"Impact of Area, Production of Tea in New Planting, Yield Rates, Export and Quantity Retained for Domestic Consumption." *TAI Newsletter*, Vol. 22, No. 11, November (1986).

"Efficiency of Tea Estates." *TAI Newsletter*, Vol. 23, No. 1, January (1987).

"Role of Tea Plantation in North India." *TAI Newsletter* 23 May (1987).

"Export Prospects of Indian Tea in Iran." *TAI Newsletter*, Vol. 24, No. 1, January (1988).

"Structural Changes in Tea Exports in India." *TAI Newsletter*, Vol. 24, No. 1, February (1988).

"Export Possibility of Indian Tea in Arab Republic of Egypt." *TAI Newsletter*, Vol. 24, No. 5, May (1988).

"Export Prospects of Indian Tea in Canada." *TAI Newsletter*, Vol. 24, No. 7, July (1988).

"Export Prospects of Indian Tea in Poland." *TAI Newsletter*, Vol. 24, No. 9, September (1988).

"Export Prospects of Indian Tea in Syria." *TAI Newsletter*, Vol. 24, No. 1, January (1989).

"International Tea Promotion." *TAI Newsletter*, Vol. 25, No. 5, March (1989).

"Export Prospects of Indian Tea in Libya and Jordan." *TAI Newsletter* 25 May (1989).

"Tea in South India." *TAI Newsletter*, Vol. 25, No. 7, July (1989).

"Production and Exports of Value Added Teas in India." *TAI Newsletter*, Vol. 25, No. 10, October (1989).

"Problems and Prospects of Coimbatore Tea Industry." *TAI Newsletter*, Vol. 26, No. 7, July (1990).

"Problems and Prospects of Wynaad Tea Industry." *TAI Newsletter*, Vol. 26, No. 10, October (1990).

"World Markets of Tea and its International Arrangements." *The Assam Review and Tea News*, Vol. 73, No. 11, January (1985).

"Tea Economy in South India." *The Assam Review and Tea News*, Vol. 73, No. 12, February (1985).

"Export Prospects of Tea in Singapore." *The Assam Review and Tea News,* Vol. 74, No. 5, July (1985).

"Structure of Tea Industry." *The Assam Review and Tea News,* Vol. 74, No. 6, August (1985).

The entire thesis, namely, "Structural Changes in Tea Plantation Industry" since 1958 was published by Assam Review Publishing Company in 17 installments, like, (i) October 1985, Vol.74, No.8, (ii) November 1985, Vol.74, No.9,(iii) December 1985, Vol.74, No.10, (iv) January 1986, Vol.74, No.11, (v) February 1986, Vol.74, No.12, (vi) March 1996, Vol.75, No.1, (vii) April 1996, Vol.75, No.2, (viii) May 1996, Vol.75, No.3, (ix) June 1996, Vol.75, No.4, (x) July 1996, Vol.75, No.5, (xi) August 1996, Vol.75, No.6, (xii) September 1996, Vol.75, No.7, (xiii) October 1996, Vol.75, No.8, (xiv) November 1996, Vol.75, No.9, (xv)December 1996, Vol.75, No.10, (xvi) January 1997, Vol.75, No.11 (xvii) February 1997, Vol.75, No.12.

"Export Possibility of Indian Tea in Iraq" Iraq-1. *The Assam Review and Tea News,* Vol. 77, No. 12, January (1989).

"Export Possibility of Indian Tea in Iraq" Iraq-II. *The Assam Review and Tea News,* Vol. 77, No. 12, February (1989).

"International Tea Promotion." *The Assam Review and Tea News,* Vol. 78, No. 3, May (1989).

"Tea in South American Countries." *The Assam Review and Tea News,* Vol. 78, No. 4, July (1989).

"Production and Export of Value Added Tea in India." *The Assam Review and Tea News,* Vol. 78, No. 6, August (1989).

"Export Prospects of Indian Tea in Jordan." *The Assam Review and Tea News,* Vol. 78, No. 8, October (1989).

"History of Darjeeling Tea." *The Assam Review and Tea News,* Vol. 88, No. 10, December (1999).

"Development Issues of Tea Industry." *The Assam Review and Tea News,* Vol. 88, No. 12, February (2000).

"Development Issues of Tea Industry." *The Assam Review and Tea News,* Vol. 89, No. 1, March (2000).

"Export Prospects of Tea in Canada." *Chai ki Baat,* Vol. 33, No. 1, January (1988).

"Export Prospects of Tea in Saudi Arabia." *Chai ki Baat,* Vol. 33, No. 2, February (1988).

"Export Prospects of Indian Tea in Iran." *Chai ki Baat,* Vol. 33, No. 4, April (1988).

"Export Prospects of Indian Tea in Poland." *Chai ki Baat,* Vol. 33, No. 5, May (1988).

"Structural Changes in the Tea Exports of India."*Chai ki Baat,* Vol. 33, No. 8, August (1988).

"Export Possibility of Indian Tea in Arab Republic of Egypt." *Chai ki Baat*, Vol. 34, No. 7, January (1989).

"Availability of Tea in Internal Market or Maximization of Export Earnings." *Chai ki Baat*, March (1989).

"Export Prospect of Tea in Iraq." *Chai ki Baat*, Vol. 34, No. 5, May (1989).

"Market Research on Tea Consumption." *Chai ki Baat*, Vol. 34, No. 9, August (1989).

"Market Research on Tea Consumption." *Chai ki Baat*, Vol. 34, No. 10, October (1989).

"Green Tea" (Part I). *Chai ki Baat*, Vol. 37, No. 3, March (1992).

"Green Tea" (Part II). *Chai ki Baat*, Vol. 37, No. 3, April (1992).

Reports and Documents

Annual Reports, Various Years, Ministry of Agriculture and Cooperation, Govt. of India.

Annual Reports, Various Years, Ministry of Finance, Govt. of India.

Annual Report, Various Years, Ministry of Rural Development, Govt. of India.

Annual Report, Reserve Bank of India, Various Years.

Annual Report, National Bank for Agriculture and Rural Development (NABARD), Various Years.

CMIE, Various Issues

NABARD Publications –

(i) Micro-finance, (ii) Watershed Development, (iii) Rural Infrastructure Development Fund, (iv) Natural Resource Management, (v) Waste Land Development, (vi) State Focus Papers, (vii) Nabard News Letter, (viii) Potential Linked Credit Plan, (ix) Occasional Papers, (x) 25 Years of Dedication to Rural Prosperity, (xi) Ware Housing Scheme, (xi) Various Funds of NABARD.

Index